Pum 1. 20

Charley

In Plain Sight:
Seeing God's Signature throughout Creation

Charles R. Gordon, M.D.

Acknowledgements

This book would not be were it not for the tireless efforts of Heather Easterday. Without her vision, persistence, and attention to the myriad details that accompany this project, it would not have been remotely possible. Thank you, Heather, for helping me set up the original designedonpurpose.com website and for sifting and sorting, for editing, and obtaining photographic permission for the wonderful art that is really the soul of this work.

Thanks also to Mary Ann Lackland at Fluency for her editorial skills and patience.

I'm grateful to Abby Gordon for contributing the content for the section on "A Caring Creator" and for her review.

Thanks also to Grant, Reid and Grace Gordon for your time and inspiration.

Finally, thanks to my parents, Sue and Brian Gordon, for pointing me in the direction of true north.

To Kimberly. Thank you for sharing your life with me.

And in honor of the memory of my good friend Reich Chandler, whose faith and courage inspire me to this day. I look forward to seeing him again.

Introduction

For since the creation of the world God's invisible qualities—his eternal power and divine nature—have been clearly seen, being understood from what has been made, so that men are without excuse.

—Romans 1:20

Something very exciting is going on in the world of science these days. Overlooked in the headlines blaring the latest bad news, we are witnessing discoveries no less earth shattering than the discovery of a new world. It's only been a few decades since James Watson and Francis Crick announced their discovery of deoxyribonucleic acid, or DNA—the stuff of genes, the code of life. Your DNA codes your appearance, your makeup, your heritage, even your temperament. And now, thanks to tireless and often unnoticed work, we have "cracked the code" and know every single sequence of DNA we carry. This groundbreaking research will help unlock the cures for diseases previously thought incurable and deepen our understanding of what it means to be human.

And there are even more insights coming from the world of science. Particle physicists are understanding matter and how it behaves in ways never before dreamed possible. Cosmologists are gaining insight on the first few nanoseconds after the creation of the universe, when physical laws were being written. Thanks to the Hubble telescope, we are able to see galaxies as they were at the dawn of creation.

Indeed, these are heady times to be a scientist! But there is also some irony here. It seems the more we know, the more we understand how little we know. The bigger the universe gets, the smaller we become. And despite (some would say because of) the marvelous advances of science, we face an uncertain future. The same science that brought the internal combustion engine also begets pollution. Germ theory brings vaccines and weaponized anthrax. Nuclear energy brings almost limitless power, as well as the bomb. The list goes on. And despite our amazing achievements, we still grapple with the deep issues of life. Why are we here? Who created us? Where is justice? What happens after we die? These questions beg answers that do not seem to be forthcoming, regardless of the amazing march of progress.

It seems to me that the debate comes down to a simple, but not easy, choice. In one camp, we are told that we are the products of completely uncaring random events. Therefore, any philosophy of life, to quote Bertrand Russell, "must be built on a firm foundation of unyielding despair." The other camp teaches that our lives are not cosmic accidents. We are fashioned by a Creator for a reason. Jesus of Nazareth teaches that we are valuable (Matthew 6:26) and that our treasure is in heaven (v.20). Either Mr. Russell is correct, or the carpenter from Galilee is on the mark. One thing is absolutely certain—there is a correct answer, and they both can't be right! And, by the way, choosing which camp you fall into will affect the course of your life and eternity.

Let me say candidly that I still sometimes wrestle with the questions. In fact, this book is dedicated in memory of my friend Reich Chandler who left us too soon. However, Reich never doubted where he was going, and his faith shone like a beacon to all who follow. Sometimes the harsh realities of life make me sympathetic to Mr. Russells's point of view. There are difficult questions posed when we are confronted with evil, when we come face to face with seemingly uncaring chance.

You would think that if Mr. Russell and his band are wrong, there would be better answers for the tough questions. But this is where I think we have a clue. If our lives really are a result of billions of random accidents with no purpose and meaning, why is the premature loss of a friend so disturbing? Shouldn't we expect it after all? If there is no justice, then why do we want it so badly? If there is no Absolute Truth, then why do we seek it?

And I think there is another clue hidden in plain sight. Buried in the first chapter of Romans, the Apostle Paul tells us, "For since the creation of the world God's invisible qualities—his eternal power and divine nature—have been clearly seen, being understood from what has been made, so that men are without excuse" (1:20). Put another way, the Creator shows himself in his handiwork. You have never seen a painting that painted itself. But you can learn a little about an artist by studying his work. Every artist who has ever painted, sculpted or drawn has a unique style. Likewise, there is this sense of orderliness as we see distinct patterns repeated endlessly in all creation in the most unexpected places.

These common elements of design dawned on me slowly. And I realized them initially quite by accident—if you believe in accidents! But the more I looked, the more I saw. And the more I saw, the more wonder I experienced. Wonder, ironically, is a lost art today. We are too busy to stop and appreciate the world around us. But if we do, we become aware of something larger than ourselves and our questions. There is a majesty to creation that defies being random. There is a sweep, a scale, a beauty—a true north—that the Bertrand Russells can't really explain.

This beauty and orderliness of creation, and the wonder that is speaks to, gives me hope. Hope that Jesus is right and we are created for a purpose. Hope that one day all will make perfect sense. And it is my prayer that, through this book, you may share this sense of wonder at the beauty of all creation. And share my hope.

Dorf hinter den Bäumen, Ile de France: A painting made by Cézanne in 1879.

Straße in Chantilly: A painting made by Cézanne in 1888.

The Original Artist

What is beautiful? A visit to an art museum is a great place to observe how the first primitive efforts of artisans built a foundation for the great masters whose skills, in turn, led the way to impressionism and some of the most beloved (and valuable) works ever produced. The 20th century brought the advent of modernism, abstract painting and even avant-garde artists whose intent, it seems, is to shock and repel the viewer instead of celebrating beauty!

But one thing remains consistent. Every original artist has a unique style. Of course, any style can be copied, but what makes an artist great is to have that something. You don't have to be an art historian to tell that the two paintings here were created by the same man, Paul Cézanne. People are drawn to his originality because he was so consistent and true to his style. You can often identify the painter just by looking at the work. The same is true for all artists: songwriters, authors, sculptors and so on. They don't have to sign their work, because their signature is in the work already!

You too have a one-of-a-kind style. Whatever you call this unique imprint, you have it. It's in your voice, your handwriting, how you approach challenges and even how you think and interact with others. You are an original, and there is no other you!

And this got me to thinking. If we have this unique signature in all we do, could it be that the Creator of the universe also has a certain style? Don't get me wrong—we can't possibly compare the creation of everything and everyone to painting a picture. But I have a basic premise: a Supreme Being created everything and he has a distinctive style. Of course, some might say this is because of underlying mathematical principles. And, sure enough, there are numerical patterns present in many of the similarities illustrated in this book. We also see math at work in great art. DaVinci, Bach, Michelangelo and many others all drew on numerical themes to create some of the world's most beautiful art. But the existence of mathematical laws doesn't really answer the underlying question: Where did those laws come from? And why should we expect any order or regularity in the universe anyway?

We live in an incomprehensibly amazing universe. One that defies simple explanation. And the more you look at the signature on his work, the more amazing, astounding and awesome the Original Artist becomes. So, the purpose of this book is to share some of these uncanny observations that are hidden in plain sight. To ask questions and to consider some possible answers. To reawaken you to the wonder of God's creation. I believe God gave us our sense of wonder and awe for a reason. Secular culture devalues this because it cuts right to the heart of the matter: we are not accidents.

For we are God's workmanship...

–Ephesians 2:10

How is your signature unique? How has it changed over the years?

Read 1 Corinthians 12. How would others describe what makes you "uniquely you"?

Purkinje Cell: A brain cell located in the cerebellum involved in processing impulses from the motor cortex.

Oak tree: A deciduous tree of the genus Quercus.

Seeing beyond the Ordinary

Years ago in medical school, I was so immersed in my studies that the anatomical and cellular world became as everyday to me as the scenery on my drive home. I'll never forget one night being confronted by a Purkinje Cell (a nerve cell carrying information from the cerebellum) growing on the side of the road. Of course, it wasn't that at all—it was an ordinary oak tree. I thought I'd lost it. "You need to get out more," I told myself. "Who sees a Purkinje Cell on the side of the road?"

The similarities in the design of the universe are amazing. The fact that something as microscopic as a nerve cell and as dense as a centuries-old oak tree complement each other so well speaks to the intentionality of the design. Even the thick connections in Purkinje Cells, which comprise the part of the brain that controls movement, are called an "arbor." So here's my question: are these similarities just a bizarre coincidence? When my eyes were opened to the uncanny resemblance of brain cells to oak trees, I was in awe. Could it be that they are following a common set of rules written by an omniscient Designer?

The problem is we're too busy to see signs of God's signature on creation. Although advances in science allow us to see more evidence of his handiwork, we appreciate it less. Sadly, creation has become commonplace. Being so immersed in our everyday world blinds us to beauty and wonder and all too often, the extraordinary reverts to ordinary. Let's say we could actually see a nerve cell with the naked eye. I wonder if it too wouldn't soon become humdrum. Bet it would!

Author Eugene Peterson once advanced a theory about why we don't spend more time worshiping and wondering at God's throne. It's not always so good to be rapt in wonder while at work. Put another way: they don't pay you to be in amazement! The very first chapter in the Bible indicates we were made to wonder and worship full time, but we've since told ourselves we can no longer afford that luxury. Preoccupied with the cares of the day, we don't take in the world around us with fresh eyes. What is it you are looking at every day that is evidence of our Creator's heart? Try opening your eyes in your everyday routine and look for His fingerprint. You will see it—seek and you will find.

They will be called oaks of righteousness, a planting of the LORD for the display of his splendor.

– Isaiah 61:3

What are you seeking today?

How does worship play a role in your daily workspace?

Try looking for signs of God's presence this week and journal your findings.

Tree Rings: Rings visible in the horizontal cross section through the trunk of a tree. One ring typically marks the passage of one year in the tree's life.

Star Trails: Concentric arcs caused by the rotation of our planet. The arcs are seen on film by leaving the shutter open on a camera.

Girl Scout Cookies, Tree Rings and Star Trails

What annual rituals do you observe? Rituals are a good thing and they play a prominent role in the Bible with annual feasts and annual fasts. We have even more rituals today, really. Think about it. Besides the big holiday traditions, we have tailgating and football games, the annual fair, election season and my favorite: Girl Scout cookie season. What do they put in those Thin Mints anyway? Scientists measure time with atomic clocks, which are accurate beyond comprehension. But most of us measure time by the turning of the earth: days and seasons. Nature is keeping time, too. Trees mark the seasons in the rings of their trunks; the sky marks time with the stars. If you leave a camera lens open all night and stare at the night sky, you'll see the stars above appear to move in similar circles. We are the ones moving, of course, not the stars. However, the concentric rotation makes the circles seem to travel in perfects arcs, much like the rings of the tree. These beautiful strokes across the sky are called star trails.

Of course, there is no connection between the two, or is there? Both measure time. Both are natural phenomena. And there is no escaping their almost eerie resemblance. These natural wonders are not obvious to the casual observer. It takes a long exposure with a camera lens to see the rings in the sky. The similarity with tree rings is amazing, but is only there for those willing to slow down and look.

Perhaps these patterns serve to remind us: God changes us over time, not overnight. Each tree ring takes a year to form, marking the slow, deliberate pace of time. It can't be rushed. Likewise, the divine process of transforming into the image of His Son, Jesus Christ, is a process.

Life itself is a marathon, though we sometimes live as though we were running a 50-yard dash toward all the things we want to do and be. Instead of instantaneous transformation, God works in seasons along the time-sensitive pattern we see in nature. He gives us a divinely determined timeframe of 24 hours a day, 365 days a year. Not a minute more. Not a day less. For Him, it's plenty of time to do everything He wants to accomplish through and in us.

How do you see your time these days? Is it rushing by you at the speed of light? Or crawling at an almost unbearable rate? Starting today, commit your whole day to God, every second and every minute. How about right now? No matter how little (or much) time is left in the day today—turn over every remaining hour to him for his divine use. Find some time to meditate on his Word. Soon you'll see the evidence that he is steadily at work in your life.

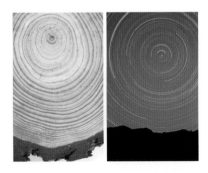

But I trust in you, O LORD; I say, "You are my God." My times are in your hands...

–Psalm 31:14–15

What annual rituals or traditions does your family have?

Why do you think God commanded the Israelites to observe numerous holy days?

Read Luke 22:7–20. What is meaningful to you about sharing the Lord's Supper?

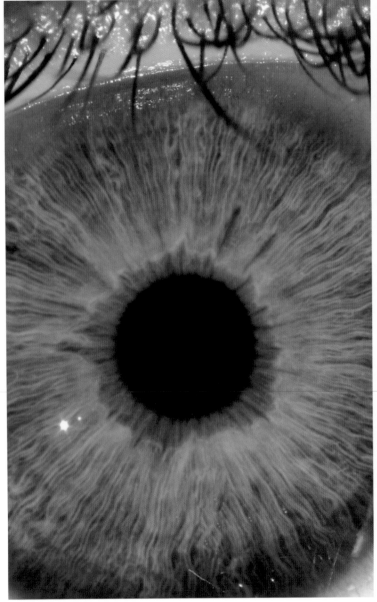

Human iris: The most visible part of the eye of vertebrates.

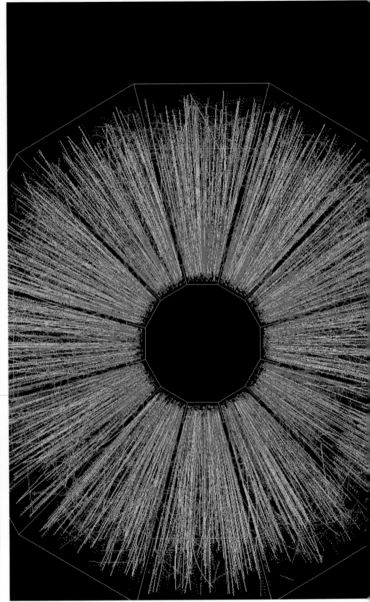

Gold Particles: The collision of gold nuclei in a particle accelerator.

We Will Be Changed

DAY 04

Some fundamental research in the field of particle acceleration is helping scientists understand the state of the universe at its conception. What they are finding is that certain materials behave differently than they previously expected. Basic elements once thought to be immutable (like gold, for instance) can and do change entirely under certain conditions. Just look at the pictures...does that look anything like gold to you? In the Middle Ages, alchemists tried to turn iron into gold. The particle accelerators of today are like alchemy in reverse. They smash gold ions together almost at the speed of light and turn gold into quark-gluon plasma. Your tax dollars at work!

The implication of this research is mind-boggling: everything that exists is subject to radical change. This sort of elemental transformation was happening when the universe began, and it continues to happen today.

These scientific discoveries remind us of what most already know: nothing stays the same in life. Everything changes: jobs, schools, hairstyles and our bodies (don't get me started). We move across the street. Across town. And across the country. We'd prefer to keep our lives stable, but there's no real possibility of being able to do that for long. Like it or not, change is an essential ingredient of the stuff of life. To live is to change.

And who better to predict change than the author of design himself? We cannot fully understand the mysteries of the world into which we are born, and so the Bible uses metaphor, as the best teachers often do. The Bible speaks metaphorically but unmistakably about a radical physical and spiritual change that is coming when Jesus returns. For those of us who know Christ personally, that's one change we will welcome with open arms.

The irony is that this will happen in the twinkling of an eye, the Bible says. The similarity of these two images obtained from a collision of gold nuclei and the transverse muscles of the iris amazes me! And they remind me to be ready for the most unexpected change of all: Christ's return, like a thief in the night.

It was not too long ago that chemists pronounced that fundamental elements, like gold, could not change. The physicists at particle accelerator labs smashed that notion to bits. And you know what? There are skeptics today that doubt Christ will return. But when he does, everything will change. Are you ready?

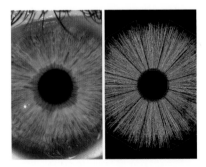

Listen, I tell you a mystery... we will all be changed—In a flash, in the twinkling of an eye, at the last trumpet.

For the trumpet will sound, the dead will be raised imperishable, and we will be changed.

–1 Corinthians 15:51-52

How do you approach change? Do you embrace it, or are you more likely to postpone it?

Read John 14:2–3. How does this verse help you have an "eternal perspective" on life?

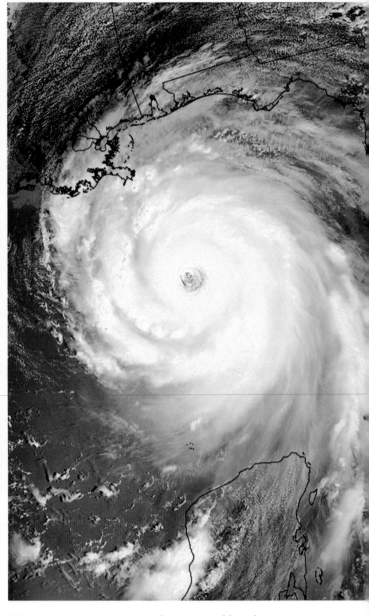

Spiral Galaxy: Cataloged as NGC 1309, one of approximately 200 galaxies that make up the Eridanus group of galaxies.

Hurricane: A storm system characterized by a low pressure system center and thunderstorms that produces strong wind and flooding ra[...]

Order and the Golden Rule

No one disputes the idea that there is order in the universe. Oh, people may claim they don't believe or that they "don't trust anything," but just watch their behavior. They turn on the light switch and expect the law of electricity to be operative. They count on the earth's gravity to hold them securely to the planet and on and on. You get the picture. The thing about physical laws is that they work regardless of how well we understand them!

With increased scientific discoveries, we are increasingly aware that some things "just are" and have been since the beginning of time. Examples include the gravitational constant, the strength of a magnetic field, the suction power of a vacuum, the tiny forces within atoms called the strong and weak magnetic forces. All of these physical laws exist in incredibly fine-tuned balance. If any one of them were just slightly different, we would not exist. In some cases, the universe would not exist. This is called the anthropic principle, meaning the universe we live in has an undeniable bias toward supporting the existence of human life.

This one principle is so stunning that non-believers struggle with its implications. Surely, they say, the universe wasn't created billions of years ago with us in mind! If not, how could we be that lucky? So many of these physical laws line up so precisely, it's hard not to believe that the odds were stacked in our favor at the moment of creation.

What are the alternatives? Well, there is the completely unprovable hypothesis that there are an infinite number of other universes, and we just happened to win the cosmic lotto. One of my favorite alternative explanations is that we are part of an alien computer simulation. No kidding. It takes a lot of faith not to believe that the world we live in (and our lives in particular) are products of intentional design.

Just as amazing is the widespread belief that there are spiritual laws as well. Though not everyone agrees as to the specifics of these laws, there is little disagreement that ignorance of these laws does not disprove their existence. You don't have to understand gravity to know that you need to be careful and watch your step. And you don't have to know the Golden Rule to realize how you treat others really matters. Entire books have been written about these spiritual laws. However, Jesus summarized them best: Love God and love others as yourself. Spiritual laws work with the same, reliable precision and order that physical laws do. Could it be the same Author penned them both?

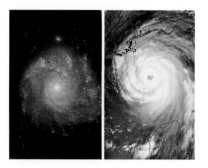

In the beginning God created the heavens and the earth.

Genesis 1:1

Where do you see order in creation?

Is there such a thing as "coincidence"?

Read Mark 6:31. Where is your "quiet place" where you go to commune with God regularly?

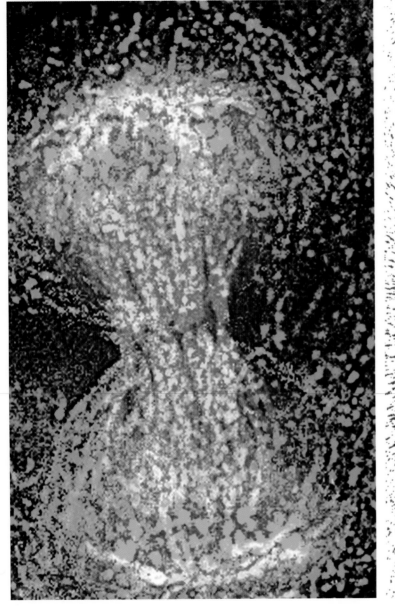

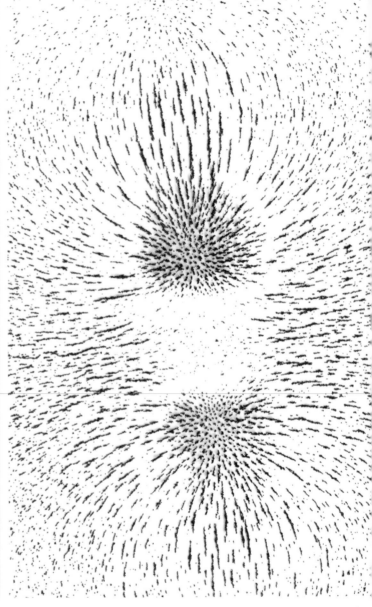

Cell Division: An epithelial cell undergoing division.

Magnetic Field: A demonstration of magnetic field lines of repelling poles using iron filings.

Order in Creation

If you think about human progress from a material point of view, you have to be amazed at the advances in the last century. We went from traveling by horse and communicating by letter carrier to commercial jets and the Internet. How did this happen? Smart people learned about the physical laws that govern our world and utilized that knowledge to make our lives better.

We were born with many of these laws already at work in our favor. For example, our planet is the perfect distance from the sun—not too close that we might burn and not too far away that we freeze. The presence of a tide-producing moon is such that it's big enough to pull tides—but not so big that we experience tidal waves all the time. We have just the right mix of gasses in the atmosphere, too. We can breathe and receive protection from UV radiation, but if the oxygen levels were too high, the planet would burn up! Too low, and we'd suffocate.

Is all this by accident? Could something so precise just happen randomly?

Consider another bedrock principle that orders our universe: the force of magnetism. Magnetism is one of those mysteries that we don't completely understand, but we see it at work in everything from the atom to the solar system. If magnetism had been a little weaker or a little stronger, it's doubtful that you would be reading this right now because, frankly, you probably would not exist.

And, as you can see in these photos, a magnetic field has arches that look strikingly similar to a cell undergoing mitosis, the process of cell division that goes on all the time in living things. It, too, follows predictable patterns governed by physical laws. The spindles of actin in the cell remind me of the same pattern in the magnetic field as they pull the cell apart. Here we have two processes—one a physical law that "just is" and another complex cellular mechanism found in all living organisms—both relying on the same underlying principles. We too depend on this highly sensitive sense of order for our own existence.

It's just another reminder that there is order in creation; there are recognizable patterns and systems in play throughout the universe. When your world seems to be falling apart, remember Who is there, holding it all together. If he ordered the universe so precisely, he can still the chaos in your life, too. Take a deep breath. And be thankful. He is in charge, so you don't have to be.

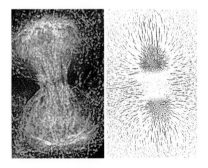

For God is not a God of disorder but of peace...

—1 Corinthians 14:33

Be thankful for the little things. What dominates your thinking—gratitude or the desire for more?

Read James 1:17. Make a list of the people, things and experiences for which you are most grateful.

Roly Poly: A member of the terrestrial crustacean group and known as Armadillidiidae. Commonly called Pill Bug or Roly Poly, due to its ability to roll into a ball.

Three-Banded Armadillo: A mammal whose name means "little armored one." It can roll itself into a ball when endangered.

A Caring Creator

Last week one of my daughters came up to her older sister to present her with what she thought was the most wonderful thing ever made. She curled up her hand and ceremoniously placed a small, dark object in her sister's palm. Upon looking at the gift more closely, her older sister saw a small "Roly Poly." These little things are also known as pill bugs or woodlice, since they eat wood. Of course, these unscientific names are not correct since they are neither bugs nor lice. They are actually crustaceans, part of the lobster family!

The moment she deposited this little creature into her big sister's hand, it rolled into a defensive ball. Thus the name, Roly Poly. With the wise authority only a three-year-old can assert, she then advised her sister not to eat it. "Never, ever, ever," she cautioned.

As my older daughter considered what to do with the gift so lovingly bestowed on her, she did what I hope we can all do more of—she took a closer look. At once, she realized that an armadillo assumes nearly the same position when frightened. The armadillo, interestingly enough, is not a member of the rodent family, contrary to what you may have heard—especially if you're a fellow Texan. They are part of the order Xnearthra; their closest relatives are the sloth and the anteater. They are in no way related to the crustacean family. Yet despite their differences, in order to escape predators, both creatures mirror the same defensive pose. Rolling into a ball is their best chance at survival.

My daughter later told me she noticed that they look something alike as well. Both creatures, though in completely different categories, have many hard, shell-like bands to protect them. She remarked later that it's just further evidence that there has to be a Creator. "And a caring one at that," she added. He cared enough about these small things to give them protection. And in so doing, he left the same fingerprint on both.

What are we to make of this? If God cares so much to provide adequate protection for a tiny crustacean and an ornery armadillo—how much more will he care for our needs? As Jesus said, "Look at the birds of the air; they do not sow or reap or store away in barns, and yet your heavenly Father feeds them. Are you not much more valuable than they?" He who placed his unmistakable fingerprint on these two creatures touched your heart as well when he lovingly made you.

"I know every bird in the mountains,
and the creatures of the field are mine."

–Psalm 50:11

Read Psalm 139. How does it make you feel to know you are "fearfully and wonderfully" made?

Read Matthew 10:29–31. Look for some sparrows! The next one you see, say a silent prayer of thanks to God for caring for you.

Nautilus Shell: The inside of shell and chambers of the Nautilus pompilius. This is a common name of marine creatures in the Cephalopod family.

Rose: a perennial flower shrub or vine of the genus Rosa.

What Beauty Teaches Us

Why would two very different creations share the same pattern? Nine hundred years ago, Leonardo Pisano (whose nickname was Fibonacci) described this spiral pattern and came up with a series of numbers that bear his name: Fibonacci numbers. Basically, you start with 0 and 1 and add the last two numbers in a sequence to form the sum of the next number. It goes like this: 0, 1, 1, 2, 3, 5, 8, 13 and so on. It turns out that this succession expands at a rate known as phi, or the golden ratio. This golden ratio was also called the divine ratio by Luca Pacioli, another Italian mathematician. Later Leonardo DaVinci referred to it as the "golden section."

What is so amazing about this ratio is how often it appears in nature—from enormous spiral galaxies, to the tiniest grains of sand. From the ratio of your fingers to each other, to fossilized shells millions of years old. This ratio is everywhere you look. Scientists have postulated reasons for its common occurrence, but no one has proven why we see this ratio so frequently.

But there is a deeper mystery here. This ratio is beautiful. Certain things—even people's faces—are intrinsically physically attractive based on this ratio. And that brings me to the real mystery. Why is there beauty at all? What Darwinian function does beauty serve? Why does it matter to any of us if something is beautiful or not?

One of my favorite places is Alaska. It truly is the last frontier, filled with enormous spaces, jagged mountains, wild bears and thousand-year-old towering glaciers. It is a place of jaw-dropping beauty and incredible danger. In fact, some of the most beautiful places on earth are also very dangerous. My friend, James, has a book called, *Death in the Grand Canyon.* Can you believe it? Hundreds of people have fallen to their deaths just trying to get a better view! Why is that? Why do we risk our lives in an effort to find beauty? Surely, there can be no evolutionary reason for this dangerous beauty-seeking behavior.

I'll tell you my theory. We are awed by beauty (and similarly repelled by the grotesque) because we are spiritual beings. Because we are more than just a bunch of chemical reactions with no meaning. Deep down within us dwells an innate appreciation for beauty—we know it when we see it.

Where do you see beauty? I encourage you to look for it actively throughout your day today, and see if you can notice more than meets the eye. Beauty was made for you—more than that, you were made for beauty. God lovingly crafted you in such a way that beauty would catch your attention and take your breath away.

By faith we understand that the universe was formed at God's command, so that what is seen was not made out of what was visible.

–Hebrews 11:3

What is beautiful to you?

Think of a time or place where you experienced the majesty of creation. Describe what you saw and how you felt.

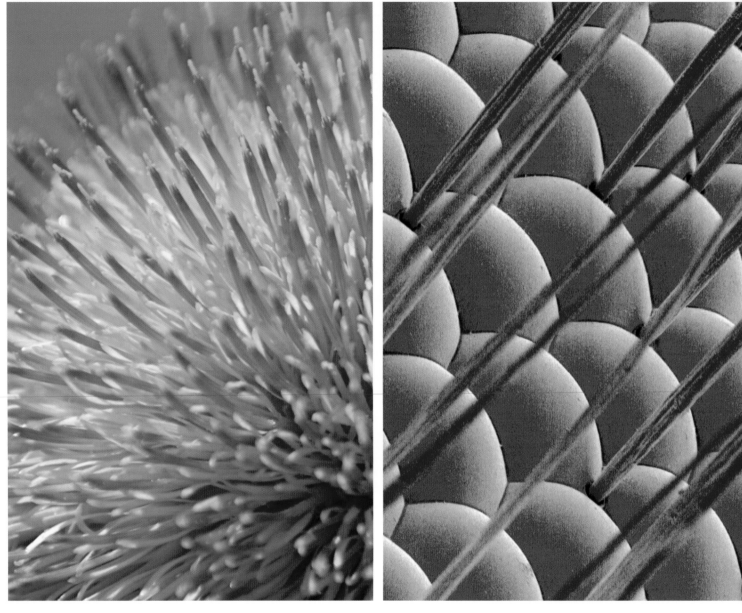

Thistle: The common name of a group of flowering plants character-ized by leaves with sharp prickles, mostly in the family Asteraceae.

Fly Eye: The compound eye of the insect with many lenses. The two spherical eyes give a fly near 360-degree vision.

Divine Purpose

One weekend, I took my two sons into the deep woods of East Texas for a manly outing of hunting and fishing. My father-in-law came along to serve as guide and soon proved what a good sport he is! As we traipsed nosily around the woods, he knew we had no chance of hurting the turkey population, but we did see some beautiful scenery.

The amount and variety of animal and plant life we witnessed in the great outdoors that weekend was staggering. A close look at the flowers blooming across the green hills reminded me of something. The small barbs emanating from some of them look just like the many antenna situated on a fly's eye.

The insect eye is marvelously designed to take in movement and color. Their eyes are compound, so they really have hundreds of lenses. Insects can also see a much broader spectrum of light than we can. When we use ultraviolet cameras to look at flowers, we can see as insects see. The feast of color in the fields lights up like billboards when viewed with these special cameras. Like a complex dance, or a beautiful symphony, all nature seems to be dancing to a tune. On our trip, the teeming hills seemed to be pulsing in an elaborate dance. There was balance, symmetry, beauty and mystery.

So why do the fly and the flower share similar design? For whom are they dancing? Why, when we really stop and look at it, do we find it so amazing? So awesome? Even the most hard-hearted skeptics have to admit the behavior of bees and flowers looks an awful lot like dancing! But who choreographed the moves? Who wrote the music?

Insects and plants don't have to wonder about their purpose. They just naturally get up and dance! They share a common Designer and instinctively join in. And though it's really beautiful, they lack what we have: free will. We can choose to join the dance or not.

You, my friend, were designed for a purpose. You know it. Deep down we know there must be a reason for our being. But what is it, exactly? We have to go beyond the bounds of science to answer questions of meaning and purpose. Throughout the Bible, we clearly see we were made for more. Our ultimate purpose is to worship God eternally. You've been invited to the dance! And the great news is that you can start dancing today.

Now it is God who has made us for this very purpose and has given us the Spirit as a deposit, guaranteeing what is to come.

—2 Corinthians 5:5

Think today about the difference between walking and dancing. Which takes more energy? Which is more fun?

Why do children seem to dance as they walk, and what does this teach you?

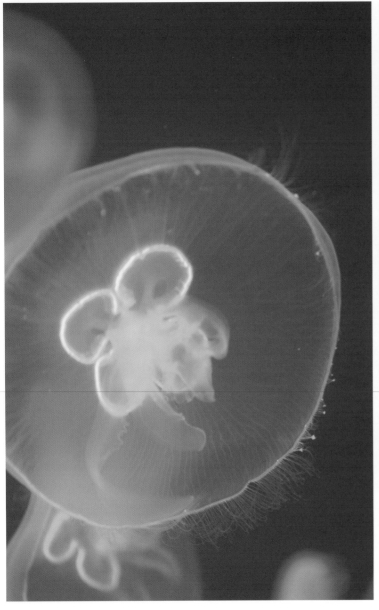

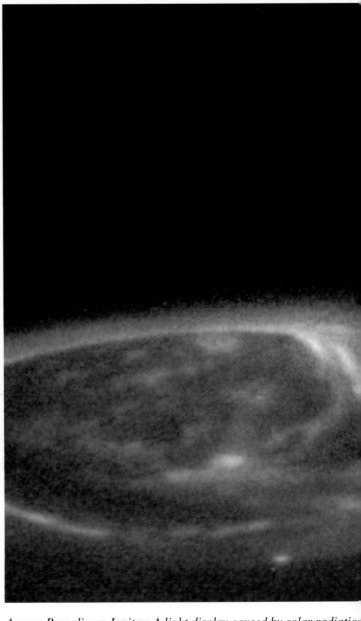

Jellyfish: Underwater shot of a Moon Jellyfish, a member of the phylum Cnidaria.

Aurora Borealis on Jupiter: A light display caused by solar radiation. On Jupiter, electric currents from its moons also cause auroras.

Worship by Faith

Ever looked in the Northern Sky late at night and seen the aurora or Northern Lights? These strange lights are caused by billions of electrons emitted from a solar storm. We can see this electric wind dancing on most of the planets. I kept looking at the images taken from the Hubble telescope and thinking that they reminded me of something. What do you know? God created two similar electric blue patterns: one he created as Jupiter's enormous aurora; the other he hid in a tiny jellyfish!

Clearly, if God wanted to make himself undeniably known, he could do it in grand aurora-like style. Instead, he is often veiled in the universe—and, like the tiny jellyfish in the sea, his purposes are often hidden to our minds. We don't always understand him. In fact, we often don't! His ways are indescribably higher—which is why we need faith.

I had a conversation with a patient the other day—a highly educated gentleman with a doctorate in theology. He mentioned he was reading a book on faith. Intrigued, I asked him why he thought God required faith. To my surprise, he responded that the more he learns about God, the less sure he is of what he believes. He said even though he used to be a chaplain in the Armed Forces, he no longer feels qualified to teach Sunday school. The more he learns, the less faith he has. It was one of the saddest and somewhat disturbing conversations I've had.

Later that evening, I sought the opinion of one of the greatest saints of faith I know: my wife. "Why does God seem hidden?" I asked her. She thought for a moment, and then she wisely said that if God were to reveal himself in the fullness of his glory, we would have no choice but to fall down and worship at his feet. Suddenly our lives would be no longer our own; we would be subjects to an awesome God. When that happens, free will is gone.

If ever we saw God for who he really is, we would have no choice but to worship. In fact, Revelation 15:4 says one day "...all nations will come and worship before you, for your righteous acts have been revealed." The reason this has not yet happened is that God desires to be loved freely, rather than to be worshipped automatically. Amazingly, the worship that he could demand at any moment is not as desirable as the love his children give freely in faith. So, if you've ever wondered what God wants from you, give him the gift of worship by faith.

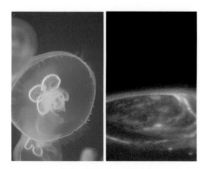

And without faith it is impossible to please God, because anyone who comes to him must believe that he exists and that he rewards those who earnestly seek him.

–Hebrews: 11:6

What does worship mean to you?

Why does God desire our worship?

Read Hebrews 13:15. How is worship good for us, too?

Eagle Nebula (also known as NGC 7293): A planetary nebula created at the end of the life of a sun-like star.

Monument Valley: The Three Sisters monument within a region of the Colorado Plateau characterized by sandstone buttes.

Are We There Yet?

When I was a kid, we used to flee the hot Texas summers and drive 12 hours to the mountains of Colorado. That is a long time to be in a car with your parents, two sisters and a German Shepherd named King. On these seemingly interminable trips, I remember how the first glimpse of a mountain peak in the distance spawned the inevitable question, "Are we there yet?" Amazingly, you can see Pike's Peak for almost 100 miles before getting near it! Even though we'd spotted the mountain, we still had to drive what seemed like forever to get to it.

Now when I look at images of the universe, I get that same are-we-there-yet feeling that I used to get as a kid in the backseat. Just look at Eagle Nebula M16, dubbed Pillars of Creation, as an example. They remind many of stalagmites in a cave, but to me they bear an almost eerie resemblance to the natural monuments in the desert, shaped by eons of wind and rain.

Look closely, because here is the amazing part: the longest cloud in the Pillars of Creation is one light year long. And they are 7000 light years away! The images we see today are not what these clouds look like now, but how they appeared when Noah was building his ark. Using our fastest rocket, it would take us 20,000 years to travel the distance of just one cloud. If you started to travel the length of the cloud when Jesus was a kid, you'd now be just over 2000 years old and only one-tenth of the way to your destination!

And it doesn't stop there.

This is just one cloud in our galaxy. The Milky Way, about 100,000 light years across, is just one of millions in the known universe. The farthest stars we can see are 13 billion light years away. Why did God make it so enormous? So mind-boggling huge? The Psalmist advises us to look to the heavens, and wow! Can you believe what's out there? Is it irreverent to ask if God is just showing off?

Astronomers are proving the Apostle Paul right—there's no way to conceive all that God has created. Are we there yet? Hardly. Our God is so grand and the scope of his work so vast we can never hope to arrive at the end of our knowledge of him. What he could only see with his naked eye caused Paul to burst into praise. Just think of all we see from the Hubble telescope! When you look up into the night sky tonight, spend some time worshipping the One who created everything you see.

However, as it is written:
"No eye has seen, no ear has heard,
no mind has conceived what God has
prepared for those who love him."

-1 Corinthians 2:9

Read Philippians 2:14, 15. What does it mean to "shine like stars"?

Make an effort this week to look at the night sky. Go to a planetarium or log on to the Hubble telescope website (www.hubblesite.org). Prepare to be amazed.

Ice in water: Ice in water generally used as a cooling agent.

Iceberg: A large piece of ice broken from a glacier or ice shelf floating in open water.

Thirsty?

Water. H2O. Agua. I remember the water fountain in the Means Memorial United Methodist Church foyer. I was just a kid, so I was glad for the standing stool in front of it. Mom used to caution me, "Don't put your lips on it!" But that's hard to do when you're learning to drink from your first fountain. It took me a while, but I finally got the hang of it.

To this day, I've never tasted more bitingly cold water. They must have cranked the cooler to just above freezing—ice would have seemed warmer! I'll never forget how hot the summer days could get in West Texas and how refreshing that water was—like an oasis on a summer day when temperatures were well over 100 degrees.

Somehow, we all seem to know instinctively how precious water is to life. "Water" is one of the first words babies learn to say. Jesus compared it to spiritual life and spoke of physical thirst in his dying words from the cross. When astronomers seek other planets that might support life, what are they primarily looking for? Water. Scientists can detect its presence by its unique thumbprint in the light spectrum. We take water for granted—it covers more than 70% of the earth's surface and makes up three quarters of our bodies.

What is it about water? Where did it come from, and why is it so special?

What I remember from chemistry is that water molecules have a peculiar makeup. The hydrogen atoms attach to the oxygen at a slightly odd angle. Electrons attached to the hydrogen atoms repel each other, while the oxygen atoms share and attract these same electrons. They are all bonded together in a delicate dance. Because of the unique properties of the water molecule, something unusual happens. The solid form of water (ice) is actually lighter than cold liquid water slightly above freezing temperature.

So what?

Well, if water behaved like almost every other substance, the solid form would sink, just like a stone. But ice floats. Put some in your glass, it works every time! Were it not for this unusual property of water, our oceans would be frozen solid, preventing the sun from penetrating the resulting miles of deep ocean ice. Our planet would be a frozen block of frost spinning lifelessly around a lonely sun in a forgotten corner of the universe.

Water has the exact properties necessary to allow ice to float and permit life to exist. What are the

odds that it would just "work out that way"? In a very real way, the next drink of water you enjoy was made with you in mind. The God of creation loves you so much that he made what you need, just the way you need it. Think about that the next time you enjoy a cool sip and thank the Maker of Water.

Jesus answered her, "If you knew the gift of God and who it is that asks you for a drink, you would have asked him and he would have given you living water."

–John 4:10

Read 4:1–42 about the woman at the well. Why did Jesus say the water he gives would permanently cure her thirst? What was her initial response?

How does Jesus satisfy your spiritual thirst?

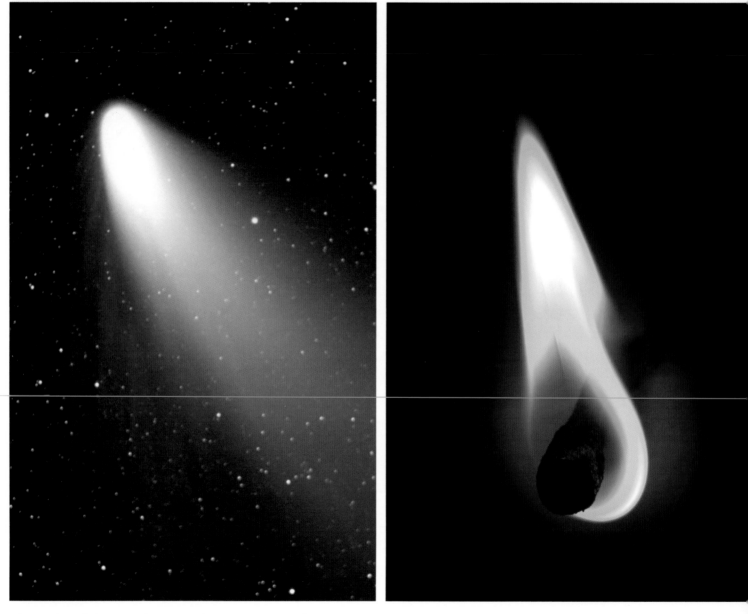

Comet: Detailed image of Comet Hale-Bopp, a celestial body observed only in the part of its orbit relatively close to the sun.

Flame: The light-emitting part of a fire.

Seen and Unseen

Recently, my children's high school basketball game went into double overtime. Although they fought a valiant effort, the Grace Cougars wound up losing by two points. During the game, I noticed people on both sides of the gym obviously praying that their side would win! (I'm not kidding!) Later that same weekend, while watching a thrilling football game between the Patriots and the Colts, I noticed one team's quarterback on the sidelines in prayer. I was certain many on the other team were praying just as fervently. With opposing and competing prayers, both teams could not win— even God could not answer conflicting prayers (at least not in the manner both sides desired). One team would win and one team would lose.

I thought of how many difficult prayers seemingly go unanswered, and my thoughts turned to my friend, Reich, whose unexpected death I mentioned in the introduction of this book. Hundreds and thousands of prayers were lifted up on his behalf, but they seemingly were not answered. Why is that?

One of the central tenets of Christianity is that there is more to this universe than what meets the eye. This world, these dimensions in time and space, are just part of God's creation. Other spiritual realms are just as real, yet unseen without the eyes of faith. One clue that tantalizes us with the notion of other realms is the existence of antimatter. Antimatter is exactly like matter, but its atoms are made of antielectrons, antineutrons and antiprotons. Comets "light up" the sky, not as a result of a flame, but because of matter and anti-matter colliding as they absorb ionic energy from the sun. The comet may look strangely like a match flame, but it is not really burning. In other words, things are not always what they seem.

The perplexing problem of unanswered prayer is counterbalanced by evidence that there is more to this universe than we can simply see, touch, feel, taste or hear. C.S. Lewis once wrote, "I believe in Christianity as I believe that the Sun has risen, not only because I see it but because by it I see everything else." There is as much to be uncovered about the mysteries of the universe as there is to know about the mystery of prayer. Are you praying about something right now? To pray is to trust. Trust God that he hears your prayer, and will answer—though perhaps not in the way you might expect.

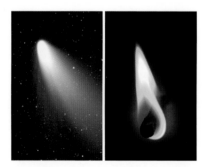

Since we consider and look not to the things that are seen, but to the things that are unseen, for the things that are visible are temporal, brief, and fleeting, but the things that are invisible are deathless and everlasting.

–2 Corinthians 4:18 (Amplified Bible)

Daily Questions

Read Luke 18:1–8. There seems to be merit in persistent prayer. About what situation have you been praying for a long time?

What answers to prayers have you seen so far?

Sunrise: The sun low on the horizon causes sunlight to pass through more air than later in the day, resulting in more molecules to scatter violet and blue light away from the eyes so more yellow, orange and red is seen.

Moonrise: The sun and moon appear the same size in the sky, but the sun is about 400 times as wide as the moon and 400 times further away.

What Beauty Says about Us

When was the last time you watched an entire sunset? I remember years ago a sunset on the beach at LaJolla, California during a family trip. It was an amazing sight. As the sun set over the Pacific, we were treated to a breathtaking display of color. From bright orange and red, to deep blue and indigo—it was dazzling. What struck me most was how everyone on the beach had the same reaction and stopped whatever they were doing. The Frisbees were stilled, and the volleyballs were silent. Almost in unison, everyone faced west, drinking in the awesome sight. Only the dogs on the beach seemed oblivious to the grandeur on display. So, here's my question: Why? When faced with beauty, why are we stopped in our tracks, mesmerized?

Once, during my neurosurgery residency, I had a long spell of working on call and went for several weeks without leaving the hospital except at night. On my first Saturday off in October, I drove through the Virginia countryside while the autumn leaves were at their peak. My eyes could almost feel the color! It was beauty so stunning that no photographer could capture.

It occurred to me later that the perception of beauty is a litmus test for mental health. If a person does not see or perceive beauty, then we call the doctors to see what's wrong. Depression, dementia, brain damage, serious psychological problems—these cause us to lose our ability to appreciate beauty. Now here is the interesting part: nobody who enjoys good mental health ever argues beauty's existence. In fact, even the most hardhearted atheist can't help appreciating a gorgeous sunset or the mountains in springtime.

So where does that innate response come from?

Beauty says just as much about us as it does about our Creator. The ability to perceive beauty (or ugliness) comes from God's Spirit infused when he made us all in his image. This awareness separates us from the animals. We know right from wrong. We recognize the difference between lovely and filthy. True beauty is universally appreciated by the mentally sound. Beauty grants us breathtaking glimpses of heaven and clues to the very heart of God.

Rarely does one think of beauty as an argument for the existence of God. But the next time someone wants to know why you believe in him, pick them a rose, or show them a sunset or point to a field of tall grass waving in the wind. Pray their eyes will be opened.

The heavens declare the glory of God; the skies proclaim the work of his hands.

Psalm 19:1

Think of the last time you saw something in creation that took your breath away with its beauty. Describe how this scene affected all your senses.

Meditate on Psalm 19.

One Grain of Sand: The tip of a spiral shell has broken off and become a grain of sand. Tumbling in the surf creates its opalescence.

Pinecone: The seed-producing cone of a pine tree.

The Miracle of Mathematics

Have you looked at all the patterns we see around us? I have previously written about the Fibonacci patterns we see in the universe. Our world is full of them—from living organisms to croissants! Anything that is coiled upon itself follows this same pattern, although you might be surprised to realize math is at work in your breakfast!

There is this crazy relationship between the natural world and mathematics. Physicist Eugene Wigner called it the "unreasonable effectiveness of mathematics." Why does gravity follow a perfect inverse square? I have a hunch: God loves numbers.

Think of the numbers in the Bible. It's impossible to read very far without running into numbers. There are three heads of the Trinity. There are seven days in the creation account. There are fifty years of jubilee in the Old Testament and 40 days and nights that Jesus fasted in the New Testament. Jesus tells us that the hairs on our head are numbered. The last book of the Bible, Revelation, is full of numbers and sequences.

God has folded numbers into the fabric of our everyday lives. There are even numbers we live by that measure the quality of our health. Numbers like our LDL, our waistline measurement and our blood pressure. Whether you know it or not, these numbers exist. And they really matter.

With the advent of computer algorithms, we are able to quantify (that is, number) things previously thought immeasurable. Not only do the pinecone and a grain of sand follow the sequence of the golden ratio, but the curves of a butterfly and the waves of the ocean also follow numeric sequences. Imagine that! Some of God's most beautiful creations boil down to a mathematic formula. Even Johann Sebastian Bach used the Fibonacci Sequence in many of his works, laboring like an architect building numerical themes throughout his music. The astounding amount of order, geometry and diversity in the universe speaks to a God of vast and unfathomable richness. For as much mystery as he left behind in creation, God also left some hefty clues as to how he did it.

When you think about it, there is no obvious reason why math should be so effective in predicting the physical world. However, physicists, astronomers, cosmologists and biologists all use inter-onnected mathematical equations, formulas, functions and relationships on a daily basis. And they all work in these widely varying contexts!

Coincidence? A mere convenience for the scientists? I think not.

Albert Einstein even argued, "The most incomprehensible thing about the universe is that it is

comprehensible." I would encourage you to put a pencil to it and find the numbers in creation. How many items can you count? What patterns do you see? And get your cholesterol checked!

*Teach us to number our days aright,
that we may gain a heart of wisdom.*

–Psalm 90:12

What is your favorite number? Why?

Read Matthew 18:21–22. Jesus spoke about numbers and forgiveness. Why did he choose such a large number in this passage?

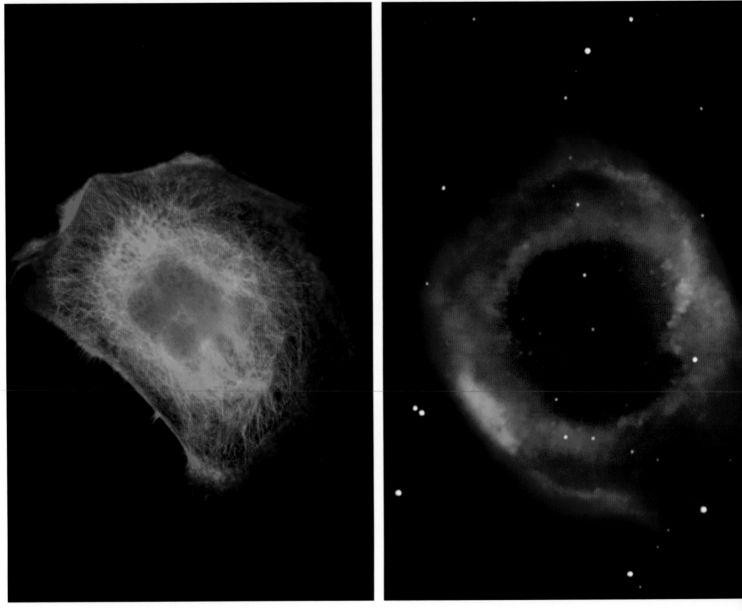

Birth of a cell: An epithelial cell during division.

Helix Nebula: A planetary nebula in Aquarius representing the blast gas and dust associated with the death of a star.

When You Want to Know Why

The Helix nebula is the closest one to our planet—a mere 450 light years away. The image we see today is actually how the star looked back when Shakespeare was writing Hamlet. The huge star is collapsing on itself, proving once more that the star—like every star including our sun—comes from perishable seed.

In contrast, our lives are more than just our short physical existence. And this is the real answer to suffering. In fact, it's the only answer that will do. Everything here is temporary. Our earthly lives are a nanosecond in the sweep of the universe and eternity. We were meant for more. For me, it doesn't require much faith to believe that God spoke and it all came to be. It isn't a stretch for me to accept that this world is ordered and that beauty exists. But what about the not-so-beautiful? How about when things go terribly wrong? Those questions are not so easy!

Looking at this problem, I have to tell you I was actually paying attention the other day to the safety demonstration on an airplane. The flight attendant said in a very serious tone, "In case of an emergency landing, leave your personal belongings behind." Really? They have to tell us this? When tragedy strikes and threatens our very lives, we don't really care too much about our "stuff."

I remember Billy Graham after 9/11 quoting Psalms and saying evil "is a mystery." And indeed it is. However, there are clues to the mystery. If there are no absolutes; if all is relative; if we are the result of a series of random senseless acts—then there is no evil. There is no right or wrong. Nothing matters in the long run, because there is no long run. But this futile point of view makes little sense to those struggling with grief and loss, because they understand life matters. Caring for grieving families is one of the toughest trials in neurosurgery. When no words are adequate, one never suggests, "Oh well, we are just a collection of chemical reactions. Cheer up—everything is pointless anyway!" Jesus said it was a good thing to mourn, because it's in mourning that we really understand how precious life is—versus a ludicrous argument that life has no meaning.

Job 5:7 tells us "man is born to trouble as surely as sparks fly upward." And no doubt, there is a lot of trouble down here. So what's the answer? How do we live in a fallen world? We can follow the Apostle Paul's example and remind ourselves, "For me to live is Christ and to die is gain," Philippians 1:21.

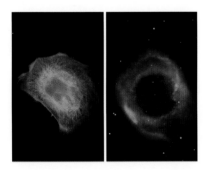

For you have been born again, not of perishable seed, but of imperishable, through the living and enduring word of God. For, "All men are like grass, and all their glory is like the flowers of the field; the grass withers and the flowers fall, but the word of the Lord stands forever."

1 Peter 1:23–26

How could the Apostle Paul believe "to die is gain"?

Read 2 Corinthians 12:1–4. If you were given a tour of heaven, how would it change your perspective on your earthly problems?

Diatom cells: Cells from microscopic algae arranged in colonies that resemble stained glass.

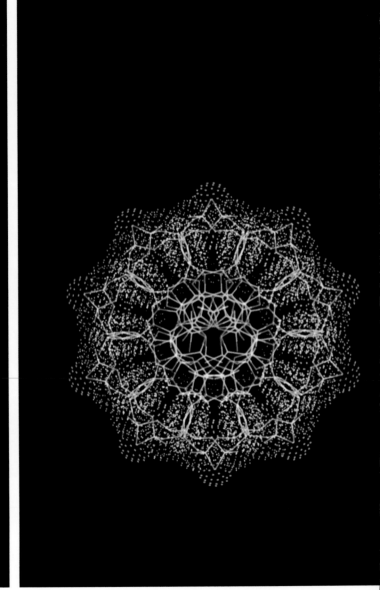

DNA strands: A cross-section of DNA as if looking down a tube ten base pairs long.

Miracles

I recently listened to a presentation by one of the greatest living scientists. Dr. Francis Collins is head of the National Institutes of Health and was head of the human genome project. In a hundred years, the significance of this project will dwarf the Apollo program. Can you imagine the implications of finally knowing every gene in the human DNA? It's like discovering a teacher's edition of your math book where all the answers are written neatly in red. Every bit as impressive is Dr. Collins' journey from atheism to faith, outlined in his book, *The Language of God*. In an age where it seems the only politically correct stance a scientist can have is that of the atheist, Dr. Collins boldly claims his faith for the entire academic world to challenge. And they do!

Dr. Collins said people sometimes chidingly pose the question, "How can a serious scientist believe in miracles?" I loved his response. It's not really a question of whether or not you believe in miracles, but do you believe in God? If you are certain there is no God, then there is no room for miracles at all. (Of course, if you are certain there is no God, you also have some pretty tough explaining to do regarding where the universe came from, why order and beauty exists— and why you are even wondering about these questions!) On the other hand, if we leave open the possibility for God as a divine all-powerful Being who created everything, including the natural laws that we observe—miracles become possible. We now have the option of believing that these laws may be briefly and rarely suspended by the law-giver.

Case in point: When Jesus was crucified, it's no wonder his disciples were holed up in a room. You don't have to take cell biology to know that people don't walk out of tombs. And there is nothing in nature that suggests otherwise. Cell biologists tell us that the minute a cell dies, vacuoles begin to form and the cell begins irreparable decay. Once the cell has lysed, it never lives again.

Or does it?

As we see in the case of Jesus' resurrection from the dead, the God who created the universe may suspend the natural order at his choosing. That's the crux of Christianity: there is not only a Creator; but he has come into our natural world and intervened. And what better sign could he give than to break the unbreakable rule and come back from the dead? It's a miracle beyond compare and worthy of our renewed sense of awe and amazement. Let that miracle come alive again in your heart today and praise God for his wondrous works.

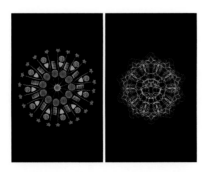

I am the resurrection and the life. He who believes in me will live, even though he dies; and whoever lives and believes in me will never die.

–John 11:25–26

Perhaps the greatest testament to Jesus' resurrection is the testimony of the disciples. Eleven of the twelve were killed for refusing to recant their belief. Would they willingly die for a statement of belief they knew was a lie?

Read John 20:24–29. We can appreciate Thomas wanting evidence before he accepted such an awesome claim. In your opinion, what is the greatest evidence of the resurrection?

Root system: All the roots of a plant form its root system.

Cellular membrane: The unit membrane of a cell. It possesses a continuous structure with one surface bordered by cytoplasm and the other by the outside world.

Watch Those Gates!

Walking on a beautifully manicured lawn one afternoon, I noticed how the blades of grass—each straining for more sunlight—looked strikingly familiar. Then it hit me: the tightly packed plants and their root system is a reflection of the cellular membrane. There again, the microscopic and everyday worlds merged in another uncanny resemblance. See for yourself!

Today, the term "grass roots" has come to mean something very different from the simple biological term. When I searched for images of grass roots online, instead of getting photos of plants, a collection of community activist groups filled my computer screen. You've got your grass roots music, grass roots social changers and grass roots gun totin' activists. But for the botanist, the phrase "grass roots" is taken literally. The root system of grass plants is an interconnected underground latticework. And what is extraordinary to the point of amazement is the similarity in appearance of my lawn to microscopic membranes.

The cellular membrane is really one of the most remarkable creations in nature. Two opposing layers of phospholipids keep the internal structures separate from the outside world. However, it's semi-permeable, so that some molecules can easily cross. Then there are gates—I'm not making this up—that allow other specialized proteins to enter and exit. These gates are triggered by molecules that recognize the proteins and allow them in, much as a sentry keeps watch. And guess what happens if the wall is disrupted? Without a good membrane, the cell cannot function and dies. In ancient cities, people paid particular attention to walls. Just like a cell, no city could long survive without the protection of a wall. One of my favorite Bible stories is about Nehemiah rebuilding the ancient wall around Jerusalem. I'm struck by the similarities between this story and the cellular membranes in our bodies. Nehemiah's job was to oversee the repair of Jerusalem's debilitated wall, without which the city was completely vulnerable. Almost as if he'd taken a cue from cellular biology, he set about rebuilding the wall—complete with gates and sentries! Not surprisingly, the neighboring enemies were displeased at his progress and tried to discourage him at every turn. Yet he refused to waiver, turning to God to save the city.

I think what inspires me most is how Nehemiah persevered in the face of adversity. What challenges do you face today? Could they be better handled if you spent some time repairing your walls, so to speak? Have the "sentries" in your life been dozing, allowing in all sorts of toxins? Amazingly, some unicellular organisms live in the harshest environments, withstanding burning heat or searing acidic environments. But they couldn't do this without strong, intact cell walls. So take a lesson from Nehemiah and the most humble protozoa—build your walls strong and watch the gates!

Above all else, guard your heart, for it is the wellspring of life.

–*Proverbs 4:23*

Read the entire book of Nehemiah as a family or with a friend—just a few paragraphs at a time. What inspires you about this story of perseverance?

What keeps you going in the face of adversity?

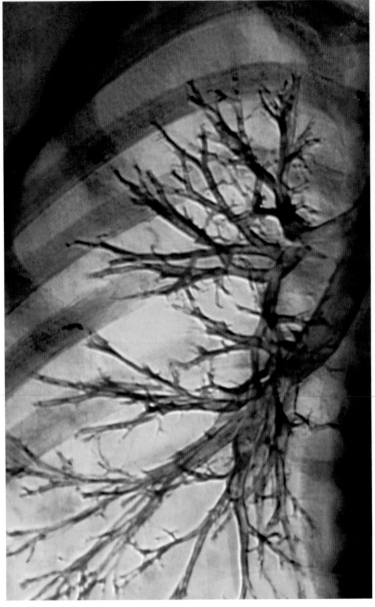

Bronchial Tree: The network of air passages that supply the lungs.

Mississippi River Delta: The result of suspended sediment deposited as the waters of the Mississippi River spill out into the Gulf of Mexico.

The Pattern of Life's Choices

DAY
19

There is something magical about branching patterns. Compare the bronchial tree of our lungs with a birds eye view of the Mississippi River delta. Each branch and tributary begins large and meanders down to the tiniest sliver. When you look at it, you can see this branching pattern everywhere: in plants, in blood vessels—even your family tree forks!

As in nature, our lives reflect much the same patterns regarding the choices we make. One of my favorite poets, Robert Frost, wrote about decisions:

Two roads diverged in a yellow wood,
And sorry I could not travel both
And be one traveler, long I stood
And looked down one as far as I could...
...I took the one less traveled by
And that has made all the difference.
–The Road Not Taken, 1916

Yogi Berra also had some classic advice: "When you come to a fork in the road, take it!" His words make us laugh, but they also make us think. So many face decisions with trepidation. Instead of boldly deciding and accepting the consequences, we wonder: Should I go this way or that way? What will the neighbors say?

We all make decisions every day, some big and others trivial. Each new choice branches off one way or the other, taking our lives in far different directions. As soon as we settle one decision, a new set of choices awaits us. And so the pattern repeats throughout our lives. I visit with patients every day facing major decisions regarding spine surgery. How can anyone know what is best? It is not always clear, and yet we must make up our minds and commit to one path or the other.

Jesus also spoke of two roads—one broad and well traveled; the other narrow and more difficult. Could it be that one single decision can affect a person's eternal destiny? He said that to choose his path and believe in him was to experience eternal life (John 17:3). When he made such an audacious claim about himself, he narrowed our choices considerably. He would either have to be deluded, despicable or Divine to say something like that. The branching pattern narrows to those just three choices at that point. No other options are available. Nor, as C.S. Lewis has pointed out, did Jesus intend for there to be!

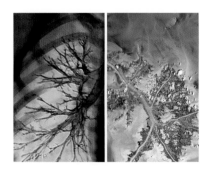

Some might say the branching similarities we see in these two pictures are complete coincidence. Or are they?

Perhaps these branching patterns are divinely placed before us as a mirror for us to understand our lives are a series of choices. None are trivial. We cannot travel two roads. Ask God for guidance on your decisions, and then commit to them wholeheartedly.

If any of you lacks wisdom, he should ask God, who gives generously to all without finding fault, and it will be given to him.

–James 1:5

How do you make decisions?

Do you go through a process or tend to go on instinct?

Read Proverbs 16:9. As you look back at important choices you've made, how have they turned out?

Wave: Spiral seen in a large hollow wave at Waimea Bay on the north shore of Oahu, Hawaii.

Fern: Spiral seen in the unfolding of a fern.

Spirals, Science and Faith

An interesting debate in *Time Magazine* once featured a couple of leading scientists. One of the speakers took the atheistic position that life simply evolved, claiming there was little chance that a God of any sort was involved.

The other took a Christian stance, noting that without certain constants of the universe being exactly tuned correctly, there would be no chance of life at all. If one looks carefully, there is unmistakable order to the universe's underlying structure. The coiled mass of a wave is the same pattern in the gentle unfolding of a fern's tiny green frond. The question is: Who put it there and why? There are so many remarkable similarities that it strikes me as uncanny and unlikely that it is random and purposeless.

In the article, I noticed the position the atheist took confused principles with application. Scientists do a wonderful job of understanding the application of elemental principles that allow us to talk on our cellphones, drive our cars, enjoy air travel, see the miracles of in vitro fertilization and even explore outer space. But what science cannot do is alter or even very well explain the underlying principles of the universe—the gravitational constant, the strong and weak magnetic forces, time itself. Science can do a good job of describing the universe, but it runs into problems explaining exactly why the universe is ordered the way it is. For example, we understand gravity works all the time throughout the universe. But we still don't understand why.

Though it is not completely silent on scientific matters, the main purpose of the Bible is to help us understand God and our relationship to each other. It is not primarily a scientific textbook, but it does outline principles of human interaction. Just as the principles of the universe can lead to scientific applications, so too can studying the basic principles of how God interacts with humans can help us understand God and ourselves.

One of the greatest foundational principles we can apply is the principle of faith. Jesus talked so often about its importance. We cannot begin to understand God or enjoy a relationship with him apart from faith. How does your faith apply in real life? Are you putting it into practice every day, or is it still just a theological principle? Faith is like a muscle. The more you use it, the stronger it becomes. And God does not ask us to have blind faith, but to open our eyes! Throughout the Bible, we are encouraged to look at the heavens and the earth as God's handiwork. We can even pray for more faith. Ask God to show you specific ways you can begin living out your faith today.

*"But let him who boasts boast about this:
that he understands and knows me,
that I am the LORD, who exercises kindness,
justice and righteousness on earth,
for in these I delight," declares the LORD.*

–Jeremiah 9:24

Daily Questions

With whom do you regularly spend time each week? How do they encourage your faith?

What exercises can you do to enlarge your "faith muscle"?

Lightning: Lightning striking the ocean.

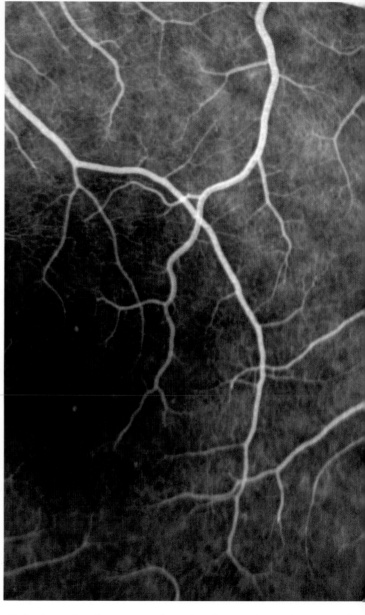

Retinal arteries: Image seen through a fluorescein angiogram as dye enters the retinal arteries.

Seeing and Believing

One of the wonderful things about nature is that it provides such a great illustration of spiritual principles. I tend to think it's because the same Creator of our physical laws also created spiritual laws. Think of some parallels. A seed is buried in the ground and rises up as a new creation. Sound familiar? The four seasons match the seasons of our lives so well, from the new birth of spring to the chill of winter. There are predators and prey, the guilty and the innocent—a seeming endless tug of war between two sides, much like the spiritual war into which we have been drafted.

And then there are storms—in nature as in life. Some are furious, like the one mentioned in Matthew, and seem to come out of nowhere. Other crises start slowly, settling in like a long winter blizzard. At times, it may seem as though the Master is sleeping and hard to rouse. "Doesn't he care?" our hearts cry out. "Jesus, I'm drowning here, don't you see?"

You've likely seen your share of storms. Jesus was no stranger to them. Every major character in the Bible faced them. Maybe you are facing one even now. How will it all turn out? Jesus gives a clue in John 16:33, "In this world you will have trouble, but take heart, I have overcome the world." Friend, there is more to this world than meets the eye. Take another look at the images: a storm over the water and the inside of the human eye. Both the storm and the ability to see it share the same Creator. However, the eye, marvelous as it is, can't see the most important things. Our eyes focus only on what's happening around us, but God looks at the heart where love, faith and hope reside.

I once had the pleasure of sitting in the co-pilot's seat of an airplane to learn about flying. I asked the pilot, "How can you fly when you can't see in a storm?" He said pilots are taught to trust their instruments, because the bad weather can play tricks on your senses. God's Word is our instrument. If you are in the midst of a storm, don't trust your instincts. Turn to your instruction manual, the Bible—and trust your instruments.

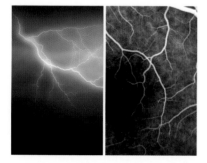

Without warning, a furious storm came up on the lake, so that the waves swept over the boat. But Jesus was sleeping.

–Matthew 8:24

Some say that in times of crisis we see who we are. What have you learned about yourself through life's storms?

Read Matthew 8:23–27 slowly. Put yourself in the disciples' boat. Feel the sting of the sea spray and the harsh wind. Where do you turn in the storms of life?

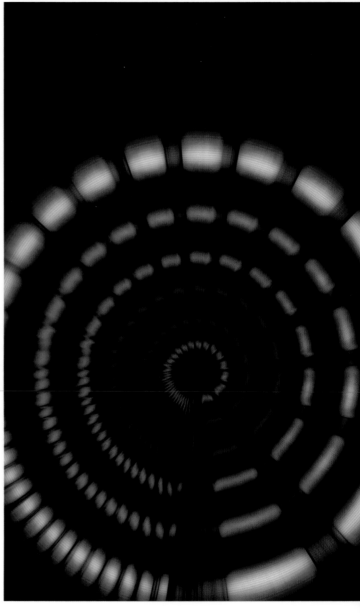

Ripples on water: Concentric circles created as momentum transfers to all the molecules of water around a stone thrown in the water.

Sound wave: Wavelet graph formed by a sparrow's song.

The Sound of Waves

There is a new field of medicine just in its infancy. The study of sleep, and the treatment of sleep disorders, is helping us understand the basic biology of the brain. It looks as if the brain must rest, and in sleep, our brainwaves change dramatically. We really do have brainwaves—electronic transmissions that can be measured. When you stop and think about waves, they are everywhere! Sound travels in waves. Light, liquids, radio frequencies, wireless Internet—all waves.

There is now widespread agreement among scientists and cosmologists of a foundational event at the start of creation. Not that long ago, the brightest secular thinkers believed the universe was static and had always existed. Not so anymore. Several strong pieces of evidence point to a defined starting point for the universe. Not only are we able to see galaxies that existed shortly after this enormous everything-out-of-nothing event, but we can still hear the radio wave echoes of the explosion. The point is this: The universe and time itself have a finite beginning. Like a pebble thrown into the water, the resulting ripples are still evident.

To Christians, the singular event of history still rippling through eternity is the resurrection of Jesus. In the last chapter of Matthew's gospel, Mary Magdalene and the other Mary head to Jesus' tomb, only to find it empty. First, they see an angel; then Jesus tells the stunned women to tell the disciples that he's alive and will see them soon. The ripples from the sound of their good news are with us still today.

Why is the testimony of these women so important? Well, if the author were going to fabricate a story that deliberately deluded readers, he would want to make it as believable as possible. The last characters he'd use to corroborate the fiction would be a couple of women, who were not at all respected 2000 years ago! Back then, the testimony of women was not even admissible in court. If Matthew were inventing his story, he certainly would not have used women to be the first witnesses. But he wrote it anyway and recorded that Jesus first appeared to the women. I can almost see him rolling his eyes and thinking, "Why did it have to be the Marys? No one will ever believe this!" But he did write it that way. And there is only one plausible explanation: because that is the way it actually happened!

I think Jesus had to chuckle to himself sending those who had no clout to announce the most amazing miracle of all. This observation is one of the most compelling to the authenticity of Scripture, particularly the resurrection. Because Jesus chose the least likely means of sharing the news of his triumph over death, you can be reassured now two millennia later that the story is true.

If there is no resurrection of the dead, then not even Christ has been raised. And if Christ has not been raised, our preaching is useless and so is your faith.

–1 Corinthians 15:13–14

What other scriptures suggest the resurrection is a historically accurate account?

Why is the resurrection the key to the authenticity of Christianity?

Read 1 Corinthians 15:3–7 regarding Jesus appearing to 500 people at the same time after his resurrection. What is the significance of the fact that when this was written, many of those eyewitnesses were still alive?

Sea Anemone: Pink Anemone photographed on the Barrier Reefs surrounding Belize.

Cilia: Ciliated cells on the tracheal epithelium surface.

It's a Small World After All

Most people playing at the beach don't think about the wide variety of creatures living in the depths of the ocean. We tend to appreciate only what we can see from our vantage point, like fish and the occasional whale or porpoise. We often forget there is an entire world of activity taking place below the waves! It's filled with a vast variety of creatures including the flower-like sea anemone, whose tentacles bear striking resemblance to microscopic cilia in our body.

However, there is an even smaller, more densely populated world of sea creatures that's impossible to see with the naked eye. And would you believe there are people who devote their entire lives to studying these miniature ecosystems of sea plants and animals? The scientist, aided by a microscope, is able to see thousands of tiny organisms, bacteria and protozoa in just one cubic foot of murky ocean water. Just think of it: When you go to the beach, or a pond, there is an entire, separate ecosystem hidden in plain sight.

Just as there is an unseen, unrecognized microphysical world right at our feet, there is also an unseen spiritual world around us as well. The Bible describes another dimension that exists alongside our material world, but it is comprised of supernatural spirit beings that are in constant battle with each other. The Apostle Paul is simply reminding us there is more than meets the eye in our world today. If you watch the news at all, the evidence is so apparent of this great supernatural clash between good and evil, right and wrong, light and darkness that few people argue about its existence. We see evidence of the struggle; we just can't see the invisible battles taking place with our eyes.

By faith, we know this spiritual realm of good and evil it is just as real as the physical world in which we live and breathe. This seemingly endless struggle between good and evil is called spiritual warfare. The Bible teaches us that our struggle is not against what we can see and touch—it's against the invisible "spiritual forces" of evil. Unlike the micro-ecosystem of the sea, this unseen realm directly affects our everyday lives. There is no neutral ground in this battle—and no room for spectators. You must choose sides and take a stand.

For our struggle is not against flesh and blood, but against the rulers, against the authorities, against the powers of this dark world and against the spiritual forces of evil in the heavenly realms.

–Ephesians 6:12

Where do you experience spiritual warfare taking place?

Why do we struggle to choose right and wrong?

Read about the Apostle Paul's struggle in Romans 7:15–25.

Cyprus Branch: Plants in the cypress family Cupressaceae, a conifer of northern temperate regions.

Green Crystal Growth: Pyromorphite Crystals on Limonite.

Colorblind

Green is my daughter's favorite. She loves everything green: grass, trees, even little green bugs. It occurred to me the other day that if a child were born colorblind, and saw the whole world in black and white, she would have a different perspective. She might ask you to describe the color green. You could give her examples of things that are green, like plants, emeralds and the Wicked Witch of the West. You could go the scientific route and say green is the light waves that come in at 480 nanometers. Or you could talk about Kermit the frog, money and even being green with envy.

But you still haven't described green. That child would not understand what these things have in common because she cannot relate to the experience of seeing green. The only way to know green is to experience it.

And as I was thinking about green, I realized that we are all colorblind. People who do not believe in God often say something along the lines of, "If God is good, why does he allow bad things to happen?" We have all experienced tragedy or misfortune that occurred to us or to those we love. It seems unfair, inexplicable and just plain wrong, and we rail against these injustices. Sadly, many then conclude that either God does not exist...or if he does exist, he does not care. Like a child born colorblind trying to envision the color green, we often do not understand what we see. We too are colorblind.

We cannot understand the scope of the universe, the link to infinity, the existence of heaven or the spiritual world. The things that God inhabits include not only the world we see, but also other dimensions. Time does not exist for him as it does for you and me. He sees time from the beginning to the end. He made time and is the author of infinity. He is in all things. The spiritual dimension defies materialistic description and understanding, no matter how hard we may try.

While God has left us clues in his creation, we will ultimately never see "in color" this side of heaven. That is why the Apostle Paul assures us that one day we will know fully. He makes the analogy of seeing a reflection in the mirror and seeing things dimly that will one day be completely visible. If you find yourself in the midst of adversity today, remember that there is more to life than what meets the eye. The God who calmed the seas holds you in the palm of his hand. Keep praying. And remember—though we are colorblind here on earth, one day we'll see clearly in eternity.

Now we see but a poor reflection as in a mirror; then we shall see face to face. Now I know in part; then I shall know fully, even as I am fully known.

–1 Corinthians 13:12

Read Romans 8:28. Can you think of a time when God worked for your good through a bad situation?

How might he be planning to work for good in your life now?

Rainbow: An arc of spectral colors appearing in the sky opposite the sun as a result of the refractive dispersion of sunlight in drops of rain or mist.

Rainbow Comet: Comet McNaught as seen over Australia.

Who Created All These?

There is a whole new cottage industry growing in a surprising place. Doubt is in. Disbelief is cool. Atheism is hot as the devil (no pun intended). In the bookselling business, on the lecture tour and even with kids camps, there seems to be a growing movement of unbelievers. I've read some of the books. They are written by smart people. But what I see is not really atheism, but scientism or the worship of science. Here on the altar of logic, these unbelievers fall on their knees and bow to science.

I want to be very clear here, because I love science. And sincere scientists search for truth. Since all truth is God's truth, we believers rejoice in every discovery. But there are some fundamental limits of science, wonderful as it is. Science is a tool that helps us understand the physical world. It can't tell us much of anything about the spiritual realm. True science is a gift, not a religion. And the worship of science is much like worshipping your monkey wrench or your hairdryer.

One of the common themes encountered in anti-theistic books is the enormous amount of evil done in the name of religion. And, no doubt, they have a point. However, even this criticism has some major flaws. First, they conveniently sweep all religions into the same basket. We have witnessed and will continue to witness human violence in the name of religion, but most proponents of faith do not subscribe to evil behavior. (Jesus in particular teaches "turning the other cheek" and to forgive "seventy-times-seven.") In contrast, the records of atheist regimes such as the Nazis or Khmer Rouge are by far the worst in modern history.

And these critics unwittingly fall into a trap when they take a stand that some behavior is wrong. Their own criticism argues for the existence of a sovereign God. Follow me here. We agree that the killing of the innocent in the name of any religion or god is wrong. But where do they get this idea of "wrong"? If there is no God, there is no right and there is no wrong. So why does it matter? Why bother about it? We ought to be at the beach having as much fun as we can in our short, meaningless lives. Either these authors are mainly out to make a fast buck, or they haven't entirely thought through their own argument.

Isaiah encourages us to look to the heavens. What we can see now is mind-blowing: comets, meteors, black holes, quasars, exploding suns and an incomprehensible vastness. Who created it all? This is a question that begs an answer that is at once self-evident and beyond the scope of science. Who created it all? The Creator. The One who created time, light, matter. He worked. He built. He saw that it was good. And I like to think that part of the reason he made us was to share it.

Lift your eyes and look to the heavens:
Who created all these?
He who brings out the starry host one by
one, and calls them each by name.
Because of his great power and mighty
strength, not one of them is missing.

–Isaiah 40:26

Daily Questions

What would you say to someone who claimed there is neither good nor evil?

When people argue about morality, what do they use as their standard of right and wrong?

Can you think of a culture that admires treachery, deception or deceit? Why is that?

Sand Waves: Coyote Buttes located in the far southern portion of the Coxcomb Ridge in Arizona, formed by petrified sand and calcified dunes.

Foam on Water: Natural materials in water or from soil reduce the surface tension of water creating the ability to produce foam.

Designed on Purpose

Great design is like art. Think about it. There is beauty in elegant design. Names like Porsche, Gucci, Apple and Chanel represent brands built and fortunes claimed. By transforming the ordinary to the extraordinary, they elevated the mundane to an art form.

In this book, I'm attempting to show the wondrous similarity in how things appear in the design of the universe. Ocean waves resemble waves on the sand. The sand, the sea and even fields of wheat waved by the wind share a common design. We are designed on purpose, and to deny it is to go against the universe itself. But by definition, great design is limiting; you can't take a shower in you Porsche or drive your iPod to school. You will never be able to fly your toothbrush. Engineers design objects with a specific purpose in mind, and part of great design is knowing exactly what that purpose is. So, let me ask you: What is your life's purpose? What were you lovingly and expertly designed to do in your limited time on earth?

If you've stuck with me this far, I hope you buy at least a little of my premise here. My contention is that we are not accidents. We were predestined and foreknown. You were created for a purpose. God chose you before the foundation of the world. I can't say what your purpose is, but I can give you a clue as to how you can know it. There is something you do when you feel his presence and his pleasure. You do this thing and time stands still. This activity is challenging, but strangely, you are drawn to it. Sometimes I feel this way while doing surgery. It's as if my hands are guided. Time flies and only when the case is over do I realize that I'm worn out!

There is no purpose greater than relationship. For me, I find purpose at the office in my relationships with my patients and co-workers. Or playing with my kids and spending time with my wife, Kimberly. In the context of our closest relationships, we find our true purpose and calling. We talk so much about finding our destiny. And yours awaits. Spend some time today praying about your life's purpose and what you are designed to do.

In him we were also chosen, having been predestined according to the plan of him who works out everything in conformity with the purpose of his will...

How well do you know your life's purpose?

Can a person's purpose change? Why or why not?

In what activity do you feel as if you're "at your best" and you sense God's pleasure?

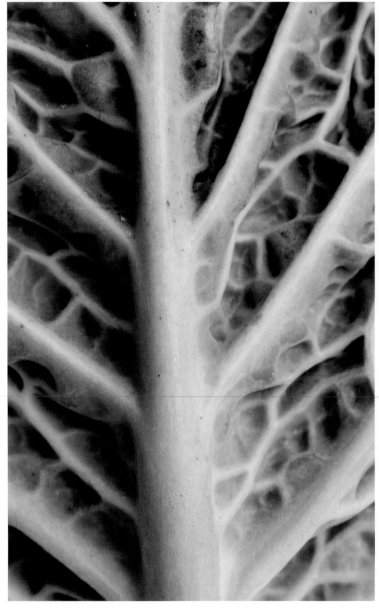

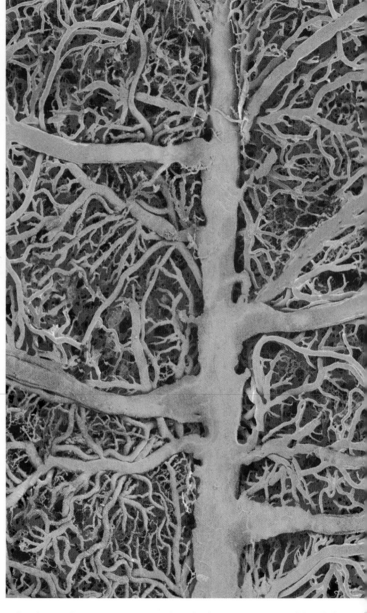

Savoy Cabbage Leaf: A leafy green vegetable and a cultivar of the species Brassica oleracea Linne of the family Brassicaceae.

Blood vessels: Intricate networks of tubes transporting blood through-out the body.

Seeking and Finding

I noticed something recently while looking through the operating microscope that made me do a double take. We were working on a part of the covering of the brain called the dura under high magnification. We could see all of the magnificent detail magnified 10 times.

And there was a leaf.

Not a real leaf, of course, but unmistakable nonetheless. As we focused in on it (and I moaned about not having a camera attachment on the microscope), we could see newly formed blood vessels where there had been stress to the body. These delicate new blood vessels were forming to help with the repair. This happens as a natural part of the healing process called neovascularization, but I had never before noticed the similarity between it and a leaf.

I've seen blood vessels hundreds of times, and the comparison is now so obvious! Why hadn't I recognized it? It makes me wonder what else I'm missing. Only when I started looking for these design "copycats" did I begin to see them. So here is my point. You are going to find whatever you seek. If you're looking and listening, you can always be inspired by nature. The problem is we're not often looking there. We're preoccupied—working, carpooling, emailing, texting—our days are filled and our attention is elsewhere. And we wonder why we can't find God.

But the amazing thing is God wants us to seek him. He wants us to find him. According to the Bible, God desires our fellowship. I have no idea why, really. But this much I know—he put a desire in our hearts to search for him. Maybe the reason why is because we are changed in the process of the search. As I've looked more intentionally for evidence of the "Creator's signature," I've become a bit different. Better, I hope. As I read and study the Bible and wrestle with the "whys," I'm less caught up in the things that don't really matter. People matter. Relationships count. Virtues mean something. Every day presents a choice: do you get caught up in the trivial irritations? Do you give into the desire to be self-absorbed and irritable? All too often I do! But there is another, better way. And it starts with seeking him.

God did this so that men would seek him and perhaps reach out for him and find him, though he is not far from each one of us.

–Acts 17:27

Read Matthew 6:21. Where is your treasure today?

Read Proverbs 4:23. How can you guard your heart? Against what?

How do you determine which positive or negative thoughts come in your mind?

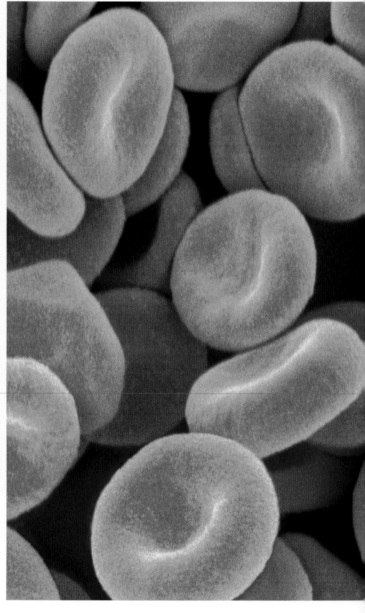

River stones: Stones smoothed through being transported by water and coming in contact with other rocks.

Red blood cells: Also known as erythrocytes, cells that deliver oxyge to the body tissues via the blood.

The Stones Cry Out

When is the last time you really looked at a rock? I confess, I've never quite understood how interesting they can be. Steve, my high school buddy, was a true rock hound—a budding geologist. He had a large collection of various crystals, igneous rocks and those weird geodes that look like an egg on the outside and have beautiful growing things inside. Today, Steve's passion has paid off as he drills for oil in the Gulf of Mexico.

I've taken another look at rocks lately. Think about how pervasive they are—they are all over the universe. Comets, asteroids and meteoroids—they're all rocks in space. What is it that scientists brought back from the moon to help them learn more about the solar system? Rocks. The Bible is full of references to stones and rocks. David gathered five smooth ones and turned a giant on his ear. The wise man built his house on one. God even refers to himself as one. We use rocks to decorate our gardens and build our homes. We drive on them in our cars, wear them as jewelry and even crush them into cosmetics. Where would we be without rocks?

Once I was looking at a creek bed and recognized smooth cells on the bottom. They weren't red blood cells, but the worn, rounded river rocks packing the stream's floor could just as well have been crowded cells in a capillary. Who crafted the rocks in the creek bed to look like those in our blood stream? Jesus said when the religious leaders asked him to keep his followers quiet that the rocks themselves would exclaim his glory. From the smallest diamond chip to the rock of Gibraltar, from the moon to the farthest galaxy—the existence of rocks speaks of a Creator who has designed and built a universe so far beyond our understanding that indeed they do cry out a thunderous chorus of praise.

The gravity of this realization should stop us in our tracks, prompting us to worship. Lest we lose our job to a bunch of lifeless stones. Wonder, you see, is the gateway to worship. We must take time to reawaken wonder and touch, taste, smell, see and hear God's creation. It's when we rid ourselves of all life's distractions and get right down to the simplicity of wonder again that we find ourselves worshipping how great and awesome our God is.

So, let me ask you again—when was the last time you really looked at a rock?

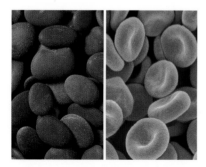

"I tell you," he replied, "if they keep quiet, the stones will cry out."

—Luke 19:40

Rocks are a symbol of security. Where do you turn for security? Your bank account? Insurance policy? Your own strength?

What is the foundation of your life? What verses from Psalms support the idea of God being the foundation of your life?

Butterfly: An insect of the order Lepidoptera. They possess the widest visual spectrum of any animal and are able to distinguish colors humans are unable to see.

Pansy Flower: Derived from hybrids of Viola species. The name pan is from the French word pensie, meaning thought or remembrance.

Who Made the Butterfly Beautiful?

When I started collecting evidence for design similarities throughout creation, I carried a little digital camera. As I looked closer at an object that caught my attention, I often spotted the little "something else" that was in plain view. Walking into the hospital the other day, I saw some flowers that I had seen many times before. Only this time I was aware of something I'd never realized. There was symmetry. And, yes—a butterfly! I hurriedly got out my camera and put my novice photography skills to work.

We are instinctively drawn to symmetry. Babies are drawn to symmetrical objects, and some would speculate this inborn recognition helps babies focus on their mothers' faces. Whatever the reason, there is beauty in symmetry and we are born recognizing it.

When you actually start to look for symmetry, it's everywhere, not just in your mom's smile. We see symmetry in the spiral galaxies and in the shape of a snowflake. It's in the poles of the earth, the butterfly's wings, the whale's tail, our two hands, the four legs of a dog, the nostrils, the eyes, the leaves on a stem, a bird's feathers, the bees' knees. It is inside our bodies, too: the two hemispheres of the brain, the vertebrae, the four pillars of the spinal column, the kidneys, all the way down to the mitotic cells and achiral carbon molecules. So why do we find this symmetry, this evidence of design, everywhere we look? Maybe a better question is: Why doesn't it amaze us more? It's so easily taken for granted.

I'm finding that if I stop for a minute, refusing to take beauty and blessings as a given, then some part of me changes a little. That's where the wonder comes in. The more I look, the more amazed I am. And the more amazed I become, the less entitlement I demand, and I realize I have been so blessed. The sun provides all the energy I need. I have yet to receive an oxygen bill from the plant kingdom. I have my parents, family and friends who pour their wisdom and love into my life.

It's like the Heisenberg uncertainty principle, but in reverse. In 1927, Werner Heisenberg said, "The more precisely the position is determined, the less precisely the momentum is known in this instant, and vice versa." What this means is that the very act of observing something changes it. But what I have learned is that the act of observation also changes the observer. There is no denying that this is a fallen world and a dangerous planet. However, we get to choose each day what we will focus on and observe. Whatever you choose will affect you. So be careful, little eyes, what you see!

The God who made the world and everything in it is the Lord of heaven and earth...

–Acts 17:24

Daily Questions

What ratio of time do you spend in worship, compared to everything else in your day?

How can you increase time for wonder and worship in your life?

Who is the happiest person you know? Why?

Fingerprint: A print made by an impression of the ridges in the skin of a finger.

Zebra stripes: Distinctive white and black pattern of African equids. The stripes help disguise zebras from predators.

Unity in Diversity

I once had the privilege of meeting with one of the world's leading spine surgeons to talk about a research study he's completing. We met at his office at UCLA Medical Center, and as I walked the halls of the hospital, I had to stop and look at a worn, black and white picture. The picture, enlarged in a nice frame, was of the founding medical staff of the prestigious hospital. All the physicians in the picture looked very serious, learned and most respectable. Just guessing, I'd say it was taken sometime in the early thirties.

But what really caught my eye was a note beside the picture, scrawled on a sticky note. It said, "If you know any of these people, call Marge at extension 355." It seems the hospital's historian was trying to match the names of the founders with the faces on the picture! Intrigued, I paused to look more carefully. There they were: bold, learned men of vision and action. Just the sort of men you hope your boys grow up to be. Think of their combined wisdom and courage! Yet most of them were begging for a name. In just three or four generations, those walking the halls of the very institution they founded did not even remember their names.

However, I realized that Someone knew their names. And the number of hairs on their heads. Someone created each one with the potential to have a unique relationship with him that would outlive their professional and personal accomplishments. I can tell you from the viewpoint of a surgeon that no two cases are alike. That's because no two people are alike. Our Designer has endowed us with unique traits, idiosyncrasies and genetic makeup that form the basis for our diversity. Not only us, but also every DNA-bearing creature on earth. This diversity extends from human life to snowflakes and every star in the universe! From fingerprints to zebra stripes, no two are identical.

What does it all mean? If you are hoping for all the answers, this is not your book!

All I can tell you is that the miracle in plain sight is the unity in diversity in creation. How can there be such endless variety, yet with common threads throughout? I humbly submit that the Creator, who knows each of us, delights in our uniqueness yet creates us with shared emotions, needs and desires. He created the same God-shaped void in your heart as he did in mine. One that only he can fill. No search for a God-substitute is fruitful. So, rejoice and be glad. No matter what you face today, God promises to direct your steps. Because he knows your name!

*I praise you because I am fearfully and wonderfully made;
your works are wonderful,
I know that full well.*

—Psalm 139:14

What traits do you share with your parents? How are you different?

What are you uniquely prepared to do to advance God's kingdom?

Shell: The shell of a member of the molluscan class Gastropoda that have coiled shells.

Cochlea: The inner ear structure which detects pressure impulses and responds with electrical impulses which travel along the auditor nerve to the brain.

A Secret Code

Some of the most dreaded diseases and problems in neurosurgery occur at the base of the skull. This area, which acts as the floor of the brain and the ceiling of the face, is a complicated, tricky location that challenges the best of surgeons. Tumors arising here can harm cranial nerve function, impairing vision and hearing. What makes them so difficult to treat is the very nature of their location. Fortunately, the field of skull base surgery has seen significant advances over my career, and the advent of focused irradiation such as the cyber knife offers new hope to patients.

The acoustic nerve courses from the ear and, along with the facial nerve, travels through the skull base via the acoustic foramen and into the brainstem. Acoustic neuromas may indolently grow here and reach a large size before they're even recognized. The eighth cranial nerve, the acoustic nerve is always at risk of injury when treating these tumors. It carries impulses from the inner ear, which is one of the most marvelous designs ever seen. And buried here in the base of the skull is the organ of hearing: the cochlea. Both beautiful and highly functional, its impressive design looks like a shell from the sea.

The cochlea is perfectly formed to help us differentiate a whisper from a symphony and everything in between. It converts sound waves to electrical impulses and sends those impulses to the brain for further understanding. Why a seashell inside our head? Who designed that and, better yet, why?

There is clearly a mystery going on here. There is this underlying order, this unfathomable intelligence behind it all. The dozens of other examples we can see every day suggest to me that there is a code right in front of us. A secret system of clues that points to an eternal mystery, just waiting to be solved. It's no coincidence, you know—the Bible calls Jesus God's greatest mystery. We can never fully understand him anymore than an ant can try to understand New York City. But thanks to his Word, we can know him and enjoy an intimate relationship with him that never ends. Like the spiral in the seashell and cochlea, God's love for us is infinite. Nothing is able to separate us from it. If our souls were not eternal, then I suppose death could separate us. Instead, God's love goes on and on forever. Today, spend some time meditating on the miracle in the mystery.

...that they may know the mystery of God, namely, Christ, in whom are hidden all the treasures of wisdom and knowledge.

–Colossians 2:2-3

Why do we love mysteries? What draws you to the unexplained? In what way is God a mystery to you?

Have you ever felt God speak to your heart? What did he say?

Read Romans 8:38–39. How would you describe God's love in your own words?

Grapes: Fruit from cultivars of Vitis vinifera. Grapes are about 80 percent water, making them an ideal low-calorie snack.

Fat Cells: Cells that specialize in storing energy as fat, also known a adipocytes and lipocytes.

Reason for Living

There is a lot of talk about "wellness" in today's medicine. The idea is that we spend too much of our resources (read: money) on treating sick people and not enough to prevent disease in the first place. Even though I have spent my entire professional life treating people who need neurosurgery, I completely agree with the wise saying that "an ounce of prevention is worth a pound of cure."

This old saying is particularly relevant today as we struggle to combat the consequences of aging. I love my older patients who are full of life. I always ask them how they do it. Everybody has a different story, but there are a few common threads in every long and vibrant life.

Something these champions of longevity all have in common is actually something they all lack: excess fat. Please understand, I love chocolate in all its forms. And I'd love to say that I see many fat ninety-year-olds in my office, but it's just not the case. Under the microscope, fat cells (adipocytes) look amazingly similar to a cluster of plump purple grapes. Both serve a similar purpose. At the heart of a grape is a seed—most of the pulp is to nourish the fragile seed until it can take root and grow. Similarly, our adipose tissue was designed to support our bodies through the lean times. Fortunately, we in the Western world don't have many lean times, leaving most of us with too much of a good thing!

Fat cells, grapes and my vibrant older patients all have something else in common—unwavering purpose. A grape is put on this earth to do one thing and do it well. Likewise, a fat cell behaves exactly as it's supposed to do. On the other hand, we have a choice—we can live according to God's design for our lives. Or we can reject it and suffer the consequences.

My healthiest patients have chosen a clear reason for living. In other words, they know and follow their purpose, whatever that may be. Christians believe that implicit with design is purpose. Unquestionably, embracing that purpose goes a long way to living longer, healthier lives. I encourage you to look again at your purpose for living: it is there. If you don't see it easily, seek godly council and read God's instruction manual so that you too may live a long and healthy life!

"**The LORD will fulfill his purpose for me;
your love, O LORD, endures forever...**"

Psalm 138:8

Daily Questions

Read 1 Timothy 4:8. What does the Bible say about the value of physical training?

What activities do you enjoy? How do take care of your body?

How do you stay true to your life's purpose?

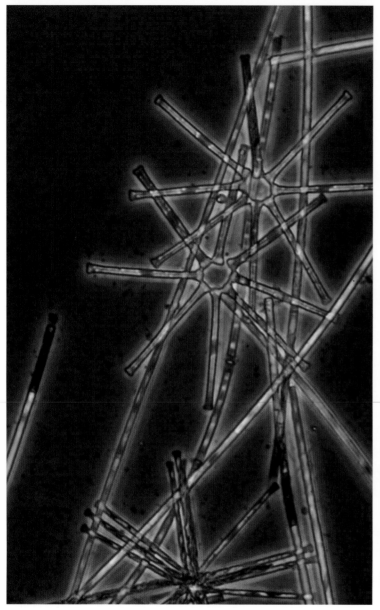

Diatom: Diatoms of Asterionella Formosa and green algae Chlorhor-midum. Most diatoms are unicellular, although they can exist as colonies.

Ice on window: Formation caused by water in the air coming in contact with and freezing on a window.

A Random World?

Where's the coldest place you've ever been? You might think about a freezing mountaintop in winter. Or a windswept prairie. But most likely, the coldest place you've ever been is on an airplane—you just didn't know it. Several thousand feet in the high atmosphere, the temperatures drop to minus 60°F and below! Just think of it—a few miles above sunny Florida, the temperature is unbearably freezing!

Cosmologists (astronomers, not cosmetologists who can give you a makeover!) tell us that a few degrees warmer or cooler and our earth would be a disaster. Either it would be permanently frozen or so unbearably hot the oceans' steam would create a permanent greenhouse effect, eventually evaporating all the water. If our earth were 1% closer or farther from the sun, it's doubtful we'd be here. If our atmosphere did not act as a heat insulator, we'd all freeze to death.

Confirming how very cold it is outside a plane, on a recent flight I noticed frost forming on the outside window of the airplane. A close-up view of frost reminds me of plankton. Our earth is so delicately balanced. Not only do we have the perfect type of sun (only about 2% of stars have the right mix of size, light and heat), but we are just the right distance from it. And that's just the start. Our orbit is almost a perfect circle so that we don't veer too close or stray too far from the sun. And our moon, much larger than most, keeps the earth from tilting on its axis as we hurl through space.

As if that was not enough, we are also in the perfect part of a spiral galaxy. Our orbit is protected from stray meteors and debris by the massive gravity of Jupiter and Saturn.

The frost melted before we landed in Dallas, and I walked out of the airport to a balmy 72-degree evening. Thankful to be a part of this remarkably proportioned planet. Some people talk about the randomness of life. As Solomon once declared, some look around and say, "Everything is meaningless." I don't buy it. Everywhere I look, I see just the opposite—overwhelming evidence of design and order, symmetry and precision. Nothing is random. God is not careless with his creation. Today, let this powerful truth remind you that you can trust him. With everything.

"Can anyone hide in secret places so that I cannot see him?" declares the Lord. "Do not I fill heaven and earth?" declares the Lord.

–Jeremiah 23:24

Where do you see evidence of a finely tuned universe?

When was a time you sensed God guiding you through circumstances?

Do you believe our lives are random? Explain.

Cerebellum: Fist-sized structure at the lower back of the brain that contributes to the motor and mental dexterity of humans.

Cauliflower: One of several vegetables in the species Brassica olera cea. It lacks the green chlorophyll found in similar vegetables becau its leaves shield the florets from the sun.

Divine Humor

My friend Stuart has such a wit. He is one of the funniest people I know. But if you don't pay attention, you'd never know it. He mutters these hysterical observations, but he never cracks a smile! I wonder if he knows how truly funny he is. I also wonder how many of his jokes I never get.

That's where cauliflower comes in. As I've said, one can find similarities of God's signature almost everywhere in creation. After I first noticed the remarkable similarity of a Purkinje cell to an oak tree while driving home from a long night of studying, I didn't expect later to see a striking resemblance of the cerebellum at the salad bar!

Don't be squeamish—the two are far different in composition and makeup, but the similar design is unmistakable, don't you think? Why is that? Is this just another wild coincidence? I'm not so sure. But I do think it's a little humorous that God would make part of our brain look so similar to a vegetable. I wonder if he was thinking something like, "Will they catch this one"? Like my friend Stuart, who rarely warns you of his dry humor, I think there is a supreme irony that part of our most complex organ looks like something you'd see on the salad bar! It's a reminder to me not to take myself too seriously.

I have to confess how fortunate I am to have a job that allows me such close contact with God's wondrous creation: his children. Every day I am privileged to work in the inner sanctum of human life. The practice of neurosurgery allows me a front row seat to the reactions and emotions of people dealing with profound physical challenges. Some handle the difficulties with grace (and even humor), others with bitterness. You know in advance that some patients are more likely to overcome their obstacles than others are. Why? It's not usually the disease of the body; it's the health of the soul that matters most.

That's why I think attitude is so important. It's not always easy to wake up and "shout for joy." The experiences of this world tend to make cynics of us all. Do you need a change of attitude today? When you're feeling down, do what the Psalmist often did—look up! Renew your perspective and be on the lookout for signs of our Designer. Start in the vegetable section!

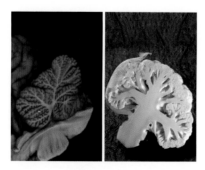

Shout for joy to the LORD, all the earth.
Worship the LORD with gladness; come
before him with joyful songs.
Know that the LORD is God.
It is he who made us, and we are his...

–Psalm 100:1–3

Read Philippians 2:3–5. Why is a person's attitude tied to his or her success in life?

How can being outside in God's creation change your perspective?

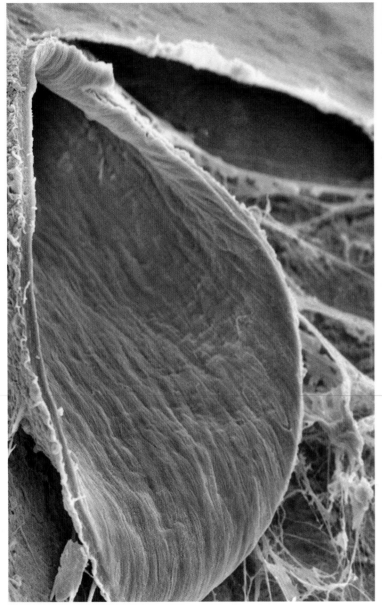

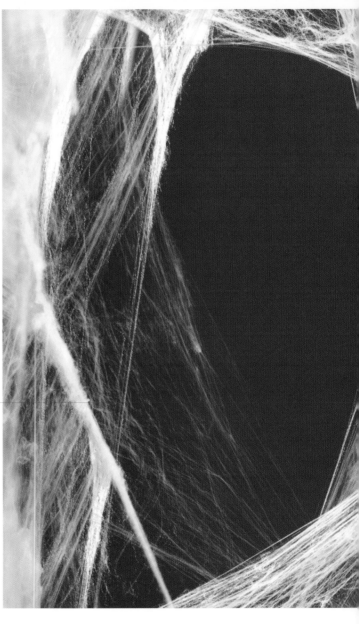

Arachnoid: A brain artery passing through the arachnoid layer of the brain. The spiderweb-like layer is one of the membranes that cover the brain and spinal cord.

Spiderweb: A device built by a spider out of spider silk extruded from its spinnerets.

A Sacred Space

I once spent a few days camping and canoeing with my youngest son and some of his best friends and their dads. We loaded up and headed to the Buffalo River wilderness in Arkansas, where there is an unspoiled wilderness and some of the best canoeing around. We camped on the riverbanks every night, and I got way up close and personal with nature! It was a great trip of fellowship and a milestone for my favorite nine-year old. I could not help but notice all the wondrous design elements all around me.

One prevalent creature around the river is spiders. One of my friends assures me that we are never more than five feet from a spider—no doubt, that was the case on this trip. What got my attention were the webs. They remind me so much of the arachnoid in the brain. See for yourself. When dissecting this stuff in the brain, it has a very fine texture. It is named after the spider's web, though its purpose and composition are completely different. This cobweb that connects all the blood vessels to each other is beautiful, nonetheless, and it is unusually strong, given its wispy-like quality and weight.

Brain surgery itself is beautiful. Whenever I operate, I am reminded that I am in a sacred space. On occasion, it is important to be able to converse with a patient undergoing brain surgery. Using special anesthesia, this is possible. And it's amazing to be looking at the part of the brain that controls speech while someone is actually speaking! It just proves to me once more that before we say anything, we think it. This is why we first need to watch what comes into our minds in order to control what we let come out of our mouths.

And the words we say really matter. The mouth is a small part of the body, but the Bible says our words control our lives much like a bridle controls a horse or a rudder controls a ship. The tongue even has the power of life and death (Proverbs 18:21). So, use your speech carefully today!

> *For I am convinced that neither death nor life, neither angels nor demons, neither the present nor the future, nor any powers, neither height nor depth, nor anything in all creation, will be able to separate us from the love of God that is in Christ Jesus our Lord.*
>
> *–Romans 8:38–39*

Daily Questions

How carefully do you choose your words?

Read Matthew 12:36–37. What importance did Jesus place on our words?

Sunflower: Annual plants native to the Americas. They possess a flowering head that is actually numerous small flowers crowded together.

Silicon Oxide Nanowires: Self-organizing nanoparticles that assembled themselves in the shape of a sunflower.

Wonder

Experience teaches us to be suspicious of something for nothing deals. Think about it. When is the last time someone offered you something of real significance for free? And, if they did, what was their motive? We say things like "there's no free lunch" and "you can't get something for nothing." And really, it's pretty much true. The Bible refers to it as the Fall. Scientists refer to it as the second law of thermodynamics. Both speak to the same truth: everything tends to run down and become a mess. This explains my garage.

The truth is we have to work just to make ends meet. The Bible does not sidestep this truth. Genesis records that as a consequence of sin, we humans were conscripted to toil by "the sweat of our brow." For our very lives depend on our efforts, and when our strength fades, so do our prospects. These facts, cold as they may seem, are borne out in our everyday experiences. The farmer doesn't reap unless he first sews and vigilantly husbands his crops. The student can't expect to excel without study. This is such a powerful reality, and it even becomes a major objection to the Gospel for some. The Bible tells us that "while we were still sinners, Christ died for us." It is unusual for someone to die for another. Such extravagant generosity flies in the face of everything we've come to know.

Children, however, don't have this problem with the Gospel. It makes perfect sense to them that they are loved and that God would sacrifice for them. After all, others feed and care for their needs every day. And here is where I think wonder comes in. Children are free to find wonder and beauty in the simplest of nature. And when we stand in wonder, in amazement, we momentarily join them. We let something in. Who made that sunset? Was it for me? The ocean spray. The rose petal. The grass. What did I do to deserve the gifts of sight and hearing?

As adults, we are so caught up in the infomaterialism of our age, using the wi-fi at Starbucks on our laptop while talking on our cellphones. We're so immersed in work, struggle and striving that we forget the gifts that we did not earn. And when we forget those gifts, we so easily become cynical. While there is plenty of room for honest doubters at Jesus' table (thanks, Thomas) you don't see any cynics. They don't accept the invitation to the banquet. Or to the dance.

But not so with you. You, friend, are a dancer! How do I know? Because a cynic wouldn't have read this far. The cynic is annoyed that this book exists and that some of us bother to spend time in wonder and awe at the majesty of creation. Of the sun with its incredible gift of (free) energy and warmth. Of self-organizing nanoparticles that fashion themselves in the shape of a sunflower. Today, let your wonder abound in childlike freedom and lavish your heavenly Father with thanks for all he has made.

Who among the gods is like you, O LORD? Who is like you— majestic in holiness, awesome in glory, working wonders?

–Exodus 15:11

Read Romans 1:20, the theme verse for this book. How do you see creation pointing to a Creator?

Why do you think wonder is the gateway to worship?

Mosquito: A common insect in the family Culicidae. Mosquitoes have mouthparts created for piercing the skin of plants and animals.

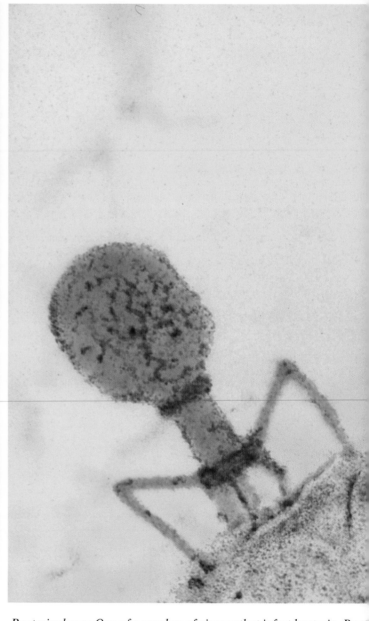

Bacteriophage: One of a number of viruses that infect bacteria. Bact riophages attach to receptors on the surface of bacteria.

Bugs, Beauty and Annoyance in Creation

In East Texas where I live, the onset of spring brings both beauty and misery. Like so much of life, along with the beautiful comes some ugly. Roses have thorns, plants have allergy-producing pollen and mosquitoes enjoy the sunny weather, too. Far beyond just being a general nuisance, they can carry deadly diseases. The Bible teaches that death, disease and suffering are a result of sin and our separation from God. I guess that means there were no mosquitoes in the Garden of Eden!

Once while trying to swat one of these pesky flying insects, I realized that I had seen its likeness in virology. The T4 Bacteriophage (killer virus) looks creepy in the same way. This virus infects bacteria by injecting its own DNA into the bacteria and hijacking the cell to become a virus-producing factory. Under a microscope, we can see one of these sinister-looking cells injecting its DNA into another cell about to become a clone-producing factory until its resources are spent and it lyses permanently. Virus comes from the Latin word meaning "poison" and no wonder!

While the virus and the mosquito are not remotely related, they have more than their menacing looks in common. Both are parasites. They can't live on their own and feed off the life of their unwilling hosts. How incredible is it then that they share so much in common! Who made the parasites? Are these God's creation as much as the beautiful butterfly? The answer is not so simple.

While we know that God created everything, the parasites by definition came later on because they could not flourish without the life of others to sustain them. Viruses cannot even reproduce on their own—they attack only good cells. C.S. Lewis once noted that there is no such thing as pure evil. It's always a distortion of what God originally designed as good. Much as cancer is a mutation of a healthy cell, and as our immune system fights these cells, the epic struggle between good and evil plays out in the human body. We see this same theme in legends, classic literature, in movies and in modern day tales. From Robin Hood, to Cinderella, to Harry Potter—every good story has a bad guy! So, pictured here are two parasitic characters in nature that resemble each other. Simply because I don't care for them so much doesn't negate their existence. The inescapable fact is that they share an astonishing resemblance: and a common Creator. Now let's be clear about one thing: just because we don't like something doesn't make it evil. Mosquitoes, virons, cancer cells—these are most unwelcome, but not evil. Evil is a term reserved for those with free will. And evil people exist, as do evil spirits. But for reasons beyond me, God tolerates them.

We know this state of affairs won't last forever. Until then, we have to trust his sovereignty. His ways are indeed higher than ours...and so is his perspective!

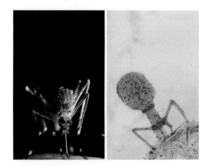

As the heavens are higher than the earth, so are my ways higher than your ways and my thoughts than your thoughts.

–Isaiah 55:9

Why is the existence of evil an obstacle to many unbelievers?

What questions do you have about why evil exists?

Read Revelation 20:7–10 about the ultimate demise of evil.
How does this passage encourage you?

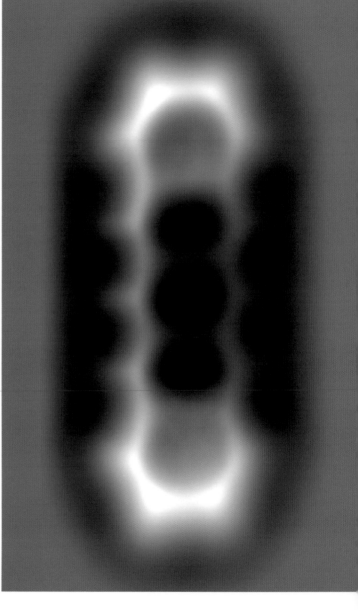

Honeycomb: Hexagonal wax cells built by honeybees in their nests to contain their larvae and stores of honey and pollen.

Molecule: Pentacene molecule imaged with an atomic force microscope. The hexagonal shapes of the five carbon rings in the pentacene molecule are clearly visible.

The Lattice of Life

When I was a kid growing up in the small town of Andrews, Texas, we had chickens in our backyard. To teach me responsibility, my mom bought the eggs for ten cents apiece, which was quite a bit more than they were worth! Now that I have a family, I sometimes wish we could raise chickens in our backyard. Sometimes.

My parents built a chicken coop out of traditional chicken wire lattice, which has a distinctive pattern of interconnected hexagons. I did not realize it then, but this pattern is the most fundamental building block of life. Carbon molecules naturally form the same hexagonal shapes that can connect like that chicken wire lattice. Carbon, the same substance in humble pencil lead and the most beautiful diamond, is crucial to life. The reason why is because the bonds carbon molecules form are uniquely suited to form the basis of organic chemistry, and even DNA itself.

And this pattern doesn't stop there. We see it repeated in honeycombs (how do bees know how to make those?) and in the covering of the eye in the cornea.

But back to the chicken coop. What did I learn in organic chemistry? Carbon forms different isomers and can support endless combinations of bonds that are the foundation of life itself. How do we know this? From direct observation. It was made that way. You can't deduce how carbon molecules bond and behave; you have to observe, measure and work hard to understand it. And although scientists may debate almost everything they study, there is no debate about the fundamental importance of carbon as the bedrock of life. If carbon molecules behaved even a tiny bit differently, none of us would be here.

Not only is the carbon molecule uniquely suited to support life, but we have an abundance of it on earth. Where did it come from? It was baked in the super-heated furnaces of stars at 100,000,000 degrees. Professor Fred Hoyle predicted in 1946 that we would find evidence of this nuclear synthesis of carbon in stars—the only place in the universe hot enough to make the stuff. And when cosmologists later looked into the heavens, there it was! Is this just another happy coincidence? Are we just fabulously lucky to live in a universe where carbon happens to work just right—and be in abundance? The Bible assures us that we were made from dust, and now we can prove that's true because that's what carbon powder is—dust!

So, it turns out that the dust our bodies are made of was formed in the super-heated furnaces of exploding stars. These stars could not have made themselves. But they were made for a reason. To make you and me. Today, you can celebrate how "fearfully and wonderfully" you are made.

The Lord looks down from heaven on the sons of men to see if there are any who understand, any who seek God.

—Psalm 14:2

When have you struggled with doubt or anxiety?

I don't know any farmers who are atheists. How much time do you spend outdoors near the source of life? How can doing so refresh your spirit?

Mt. Fuji: The highest mountain in Japan and an active volcano.

Anthill: A pile of earth built up at the entrance of the under-ground dwellings of ant colonies.

Of Mountains and Molehills

This is what I love about ground-breaking science: the better we understand the rules, the more we are able to use those rules in our favor. But you never hear a scientist saying, "Now, I'm going to make up some of my own rules!" No. Good science is about understanding the rules better so that our lives can be improved. The Wright brothers took advantage of the laws of aerodynamics, but they did not create them. You never heard Thomas Edison or Albert Einstein claim to have created new laws and principles; they made it their life's work to understand the existing laws better and harness the power of that understanding. And so it is with every scientific advance.

Truth is, we all live by inflexible physical laws every moment of our lives. However, unbreakable spiritual laws affect our lives with the same precision. This is where I think our current politically correct multiculturalism fad is unhelpful. Multiculturalism says all cultures are equal and all truth is relative. Listen and you may hear someone say, "Well, that may be true for you, but it's not true for me."

And this sounds nice, doesn't it? The problem with it is...well, it just isn't so.

Truth is not an acquired taste! And seeking the truth is not the same as just figuring out what we like. Understand—spiritual laws in the universe operate just as the physical ones do, regardless of our prevailing cultural winds.

And deep down, we know this is the case. Anyone who cares for social justice—for the suffering of the innocent and for the fate of the downtrodden—has made a stand. And we agree with those who stand for the rights of the oppressed because it is right. But you can't have it both ways. You can't say that everybody's truth is equal and then claim that it's wrong to promote child slavery, for example. Freedom of religion is one of our most cherished values. But just because a person has the right to practice his or her religion does not make it true. You also have the right to flap your arms, but that doesn't let you take flight! When we encounter spiritual laws head on, we realize that our own efforts to draw near to a holy God are not much better than flapping our arms and trying to fly.

I well remember an awkward kid back in school. He talked and dressed differently. He was socially out of place. And we made fun of him. I knew it was wrong but did not have the guts to stand up to the crowd. No culture, no talk show and no religion elevates such behavior. I broke a spiritual law, and I needed forgiveness. I still miss the mark and need forgiveness every day—just ask the people who know me best!

As it turns out, all the world's religions refer to spiritual laws and absolute truths. We just can't live up to any of them! They are too exacting. I've tried, but I just can't cut it! That's the bad news. However, the Bible reveals an awesome secret. The Writer of all true spiritual laws came to earth to offer another option. By accepting Christ and relying on his sacrifice, I can experience spiritual grace not otherwise accessible to me. Jesus did not change spiritual laws any more than the Wright Brothers changed the laws of aerodynamics. But he did give us a way of relating to God that had never before existed. And this "new way" of grace is possible—it really works! That's why they call it the Good News. Trust me, there has never been better news than this.

Picture it this way. When you look at the anthill and the volcano, you see the same basic conical shape. The combined forces of friction and gravity affect the earthen soil removed from below, which settles as a cone. But if you needed to climb thousands of feet above the valley floor, which cone would you rather climb—an anthill or a volcano? A volcano, of course!

The God of creation has set the behavior bar far higher than the tallest volcano—absolute perfection. There's no way my anthill of efforts even comes close. So God—following the very laws he wrote—brought out of Bethlehem a mountain of truth and sacrifice. His name is Jesus. Now when I build my anthill on top of his peak of perfection, I'm accepted. Regardless of how often I try and fail, I'm already on the mountaintop. The base of my tiny anthill is as high as it needs to be. Why? Because I believe. My belief puts me on the pinnacle, and because of my belief, I am forgiven! Friend, no matter what you've done, God is not mad at you. And he knows you can never build your tiny hill tall enough. He has given you a mountain so you can believe and find yourself on top of the world!

Therefore, there is now no condemnation for those who are in Christ Jesus, because through Christ Jesus the law of the Spirit of life set me free from the law of sin and death. For what the law was powerless to do in that it was weakened by the sinful nature, God did by sending his own Son in the likeness of sinful man to be a sin offering. And so he condemned sin in sinful man, in order that the righteous requirements of the law might be fully met in us, who do not live according to the sinful nature but according to the Spirit.

Those who live according to the sinful nature have their minds set on what that nature desires; but those who live in accordance with the Spirit have their minds set on what the Spirit desires. The mind of sinful man is death, but the mind controlled by the Spirit is life and peace; the sinful mind is hostile to God. It does not submit to God's law, nor can it do so. Those controlled by the sinful nature cannot please God.

You, however, are controlled not by the sinful nature but by the Spirit, if the Spirit of God lives in you. And if anyone does not have the Spirit of Christ, he does not belong to Christ. But if Christ is in you, your body is dead because of sin, yet your spirit is alive because of righteousness. And if the Spirit of him who raised Jesus from the dead is living in you, he who raised Christ from the dead will also give life to your mortal bodies through his Spirit, who lives in you.

Therefore, brothers, we have an obligation—but it is not to the sinful nature, to live according to it.

- Romans 8:1-12

Daily Questions

Why can't we earn our way to the top of "the mountain"?

Read Romans 8, a passage my pastor calls the Mt. Everest of the Bible, aloud and slowly. I encourage you to take in the view and journal all you learn about God's character from this vantage point.

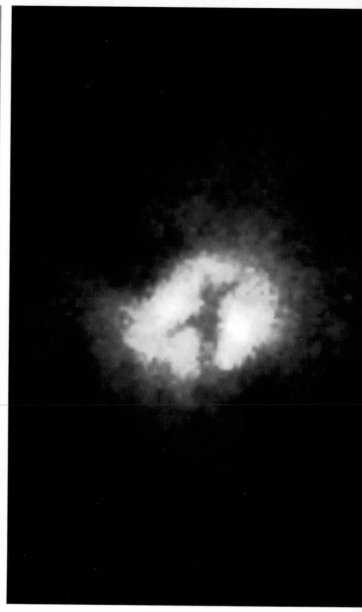

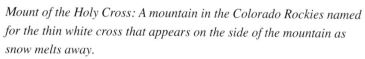

Mount of the Holy Cross: A mountain in the Colorado Rockies named for the thin white cross that appears on the side of the mountain as snow melts away.

Pinwheel Galaxy: "X" Structure at Core of Whirlpool Galaxy (M51

It is Finished

I once had the privilege of eating dinner with Dr. Paul Julienne and his wife. He is a physicist at the top of his game at the National Institute of Standards and Technology in Gaithersburg, Maryland. Dr. Julienne is interested in atomic physics and quantum theory. He told me about a quantum computer that physicists would like to build that could be vastly more powerful than any computer existing today. They already have a strontium clock that ticks something like 429,000,000,000,000 ticks per second, so you never have to be a nanosecond late again!

As Dr. Julienne looks at the universe, he also sees amazing order—an intelligence, or as he describes it, "a logos." It reminded me of the opening chapter of John's gospel. "In the beginning was the Word (Greek: logos)..." This verse takes us back before Genesis 1 when God got busy making the universe. Even before then, there was the Word. We can't conceive of a time before time. What's that look like anyway? How's that compare to say, a long weekend? Albert Einstein said that without light, there is no time; before it all began, time had no meaning.

Hang with me here.

Way before there was anything, pre-Genesis 1 or the big bang, there was something. And John calls that something logos—the Word. This Word is more than just information; we are talking about what C.S. Lewis refers to as "the deep knowledge." And then John goes a step further and says, "...and the Word was with God and the Word was God"! This Word simply spoke and set this mind-numbingly huge universe into being. Friend, there is such enormous evidence that the universe was created by an infinite intelligence that put everything in motion—it is near impossible to deny.

But what I find so amazing, completely humbling and utterly breathtaking is what John says in his next paragraph. "And the Word became flesh and made his dwelling among us," (John 1:14). This Supreme Force took the shape of a man, though still fully God, and lived on this ball of a planet for a few decades. His name is Jesus Christ.

Why did he do that? Because sin separated us from God and carried the penalty of death (Romans 6:23), he willingly sacrificed himself on the cross to bring us back to God. As Jesus declared the words, "It Is Finished," on the cross, he was not speaking about his immediate ordeal. Rather, he was declaring that he had accomplished what he set out to do in coming to earth. It was an eternal cry—claiming victory over sin and death.

This is no fairy tale, though it seems too good to be true. You simply have to respond by faith that there is a God. He knows your name and loves you more than you can imagine. He offers you complete forgiveness and the free gift of eternal life through his Son, Jesus (Ephesians 2:8-9).

"Yet to all who received him, to those who believed in his name, he gave the right to become children of God..."

– John 1:12

Read John 3:1–21. According to this passage, what does it mean to be born again? If you have not done so already, take this opportunity to express your belief in Christ.

Read Acts 8:26–39. How did this new believer publicly express his faith? If you have accepted Christ, get in a Bible-believing church and be baptized, just to show off your faith!

If you prayed to ask Jesus Christ to be your Lord and Savior as a result of reading this book, I'd like to know and celebrate your decision. Email me about your decision to follow Christ at **charley@DesignedOnPurpose.com.**

About the Author

Dr. Gordon is a physician, a scientist, an entrepreneur and a visionary. Most of all, he's a man uniquely committed to seeing God's handiwork in daily life—whether in the operating room or playing with his kids.

He is an American Board of Neurological Surgery certified neurosurgeon based in Tyler, Texas. A summa cum laude graduate of Texas Tech University, he completed his medical education at Baylor College of Medicine. His internship and residency were completed at the Medical College of Virginia. After residency, he participated in a post-graduate orthopedic spine surgery fellowship in Richmond, Virginia.

Dr. Gordon helped found Texas Spine & Joint Hospital in Tyler where he practices and serves on the board of directors (www.TSJH.org). Additionally, he serves on the medical staff at Mother Frances Hospital and East Texas Medical Center. Today his clinical efforts at Gordon Spine Associates (www.gordonspine.com) focus on helping treat people suffering from spinal disorders. He serves on the boards of Biologos, Grace Community School in Tyler and Flexuspine, a device company he founded. In addition, he serves on the editorial board of OrthoKnow, a publication serving the orthopedic and spine industry. He counts his family as his greatest gift. He and his wife Kimberly have two boys and two girls.

Email him with your own observations at observations@DesignedOnPurpose.com

About Designed on Purpose

Designed on Purpose seeks to reawaken people to the wonder and beauty of God's creation. Dr. Charley Gordon created the devotional book In Plain Sight: Seeing God's Signature throughout Creation and writes musings on his blog, www.DesignedOnPurpose.com, for fellow wonder-ers like himself who enjoy exploring the world God created and dare to ask why. Why do brain cells resemble oak trees? Why do galaxies resemble hurricanes? Why does a painter have a certain recognizable style? Designed On Purpose exists to catalogue evidence that a Creator designed the universe. The patterns show up where one would least expect, and the similarities are, well, remarkable.

www.DesignedOnPurpose.com

Image Credits

Day 1	*Images of paintings by Paul Cezanne are in the public domain*
Day 2	*Purkinje Cell - © Carolina Biological Supply Company/Phototake, Inc.*
	Oak Tree © iStockphoto.com/Maurice van der Velden
Day 3	*Tree Rings - © iStockphoto.com/Joze Pojbic*
	Star Trails - © iStockphoto.com/Reuben Heydenrych
Day 4	*Human Iris - © iStockphoto.com*
	Gold Particles - Courtesy of Brookhaven National Laboratory
Day 5	*Spiral Galaxy - NASA, ESA, The Hubble Heritage Team,*
	(STScI/AURA) and A. Riess (STScI)
	Hurricane - NASA/Jeff Schmaltz, MODIS Land Rapid Response Team
Day 6	*Cell Division - © ISM/Phototake, Inc.*
	Magnetic Field - © iStockphoto.com/Stephan Hoerold
Day 7	*Roly Poly - © iStockphoto.com/ Joseph Calev*
	Three-Banded Armadillo - © Mark Payne-Gill/naturepl.com
Day 8	*Nautilus Shell - © iStockphoto.com/David Hillerby*
	Rose - © Michelle Uhri
Day 9	*Thistle - © iStockphoto.com/Adrienne Miller*
	Fly Eye - © Susumu Nishinaga / Science Photo Library/
	Photo Researchers, Inc.
Day 10	*Jellyfish - © iStockphoto.com/Nils Kahle*
	Aurora Borealis - NASA/ESA, John Clarke (University of Michigan)
Day 11	*Eagle Nebula - NASA, ESA, STScI, J. Hester and P. Scowen*
	(Arizona State University)
	Monument Valley - © iStockphoto.com/Christopher Russell
Day 12	*Ice in water - © iStockphoto.com*
	Iceberg - © iStockphoto.com/Deborah Benbrook
Day 13	*Comet - © iStockphoto.com/T. Puerzer*
	Flame - © iStockphoto.com
Day 14	*Sunrise - © iStockphoto.com/Peeter Viisimaa*
	Moonrise - © iStockphoto.com/Neta Degany
Day 15	*Grain of Sand - © sandgrains.com*
	Pine Cone - © iStockphoto.com/Asli Orter
Day 16	*Cell - -© Jennifer Waters / Science Photo Library/*
	Photo Researchers, Inc.
	Helx Nebula - © iStockphoto.com/Manfred Konrad
Day 17	*Diatom cells -© Steve Gschmeissner / Science Photo Library/*
	Photo Researchers, Inc.
	DNA strands – © Computer Graphics Laboratory,
	University of California, San Francisco, Dr Robert Langridge
Day 18	*Root System - © iStockphoto.com*
	Cellular Membrane - © Dennis Kunkel Microscopy, Inc./
	Phototake, Inc.
Day 19	*Bronchial Tree - © Alain Pol/ISM/ Phototake, Inc.*
	Mississippi River Delta - NASA/GSFC/METI/ERSDAC/JAROS,
	and U.S./Japan ASTER Science Team
Day 20	*Wave - © iStockphoto.com/Chuck Babbitt*
	Fern - © iStockphoto.com/ Andrea Gingerich
Day 21	*Lightning - © iStockphoto.com/ Tom Hahn*
	Retinal arteries - © Thomas W. Bochow, M.D., M.P.H.,
	Eye Care associates of East Texas
Day 22	*Ripples on Water - © iStockphoto.com/Mustafa Deliormanli*
	Sound Wave - © Aguasonic Acoustics /Science Photo Library
	Photo Researchers, Inc.
Day 23	*Sea Anemone - © iStockphoto.com/Robert Dalton*
	Cilia - © Dennis Kunkel Microscopy, Inc./Phototake, Inc.
Day 24	*Cyprus Branch -© iStockphoto.com/Fedor Kondratenko*
	Green Cyrstal - © Visuals Unlimited/Corbis
Day 25	*Rainbow - © iStockphoto.com/*
	Rainbow Comet – © Mike Salway/www.mikesalway.com.au/
Day 26	*Sand Waves - © Tom Dempsey/Photoseek.com*
	Foam on Water - © iStockphoto.com/Iryna Shpulak
Day 27	*Savoy Cabbage - © iStockphoto.com/Gabriela Schaufelberger*
	Blood Vessels - © Susumu Nishinaga / Science Photo Library/
	Photo Researchers, Inc.
Day 28	*River Stones - © iStockphoto.com*
	Red Blood Cells - © MicroScan/Phototake, Inc.

Day 29   *Butterfly - © iStockphoto.com*
           Pansy Flower - © iStockphoto.com/Willi Schmitz
Day 30   *Fingerprint - © iStockphoto.com/Chris Hutchison*
           Zebra Stripes - © iStockphoto.com/Tony Campbell
Day 31   *Shell - © iStockphoto.com/Sergey Galushko*
           Cochlea - © David Spears/Last Refuge, Ltd./Phototake, Inc.
Day 32   *Grapes - © Michelle Uhri*
           Fat Cells - © Collection CNRI/Phototake, Inc.
Day 33   *Diatom - © Roland Birke/Phototake, Inc.*
           Ice on Window - © iStockphoto.com/Christopher O Driscoll
Day 34   *Cerebellum - © Dr Colin Chumbley /Science Photo Library/*
           Photo Researchers, Inc.
           Cauliflower - © iStockphoto.com/Achim Prill
Day 35   *Arachnoid - © Steve Gschmeissner /Science Photo Library/*
           Photo Researchers, Inc.
           Spiderweb - © iStockphoto.com/Joshua Blake

Day 36   *Sunflower - © Michelle Uhri*
           Silicon Oxide Nanowires -Image by S.K. Hark.
           Photo courtesy of Materials Research Society
Day 37   *Mosquito - © iStockphoto.com/Douglas Allen*
           Bacteriophage - © Biozentrum, Universtiy Of Basel /
           Science Photo Library/Photo Researchers, Inc.
Day 38   *Honeycomb - © iStockphoto.com/Oleg Prikhodko*
           Molecule - Courtesy of IBM Research – Zurich
Day 39   *Mt. Fuji - © iStockphoto.com/J Tan*
           Anthill - © iStockphoto.com/Todd Keith
Day 40   *Mount of the Holy Cross - © Jody Grigg*
           Pinwheel Galaxy - H. Ford (JHU/STScI), the Faint Object
           Spectrograph IDT, and NASA